Band 21 · August 1985

8

isotopenpraxis

AKADEMIE-VERLAG

ISSN 0021-1915 Isotopenpraxis, Berlin 21 (1985) 8, 261–296

EVP 9,– M

An unsere Autoren

Die Zeitschrift veröffentlicht Beiträge zu folgenden Themen: Kontrolle und Analyse radioaktiver und stabiler Isotope, Anwendung radioaktiver und stabiler Isotope in allen Gebieten der Forschung und Technik, der Industrie, Landwirtschaft und Medizin, Herstellung radioaktiver und stabiler Isotope sowie von markierten Verbindungen, kernelektronische Geräte für die Isotopenanwendung (Dickenmesser, Dichtemesser u. ä.) Detektoren für isotopentechnische Arbeiten und den Strahlenschutz in Isotopenlaboratorien, Meßtechnik und Strahlenschutz, Dosimetrie in Isotopenlaboratorien, kerntechnische Geräte für die Isotopenanwendung (Beton, Bestrahlungsanlagen u. ä.), Strahlenchemie, Radiochemie, ökonomische und perspektivische Fragen, Fragen der Weiterbildung.

1. Aufnahme von Manuskripten

Zur Veröffentlichung angenommen werden nur druckreife, noch nicht anderweitig publizierte Beiträge in deutscher, englischer oder russischer Sprache zum Themenbereich der Zeitschrift und zwar

Originalarbeiten
Kurze Mitteilungen
Aus der Praxis
Zusammenfassende Berichte
Tagungsberichte
Informationen.

Sind der Beitrag oder Teile daraus bereits veröffentlicht oder ist eine anderweitige Veröffentlichung vorgesehen, so muß dies der Redaktion mit genauer Angabe der Zweitveröffentlichung mitgeteilt werden. Mit der Annahme des Manuskriptes behält sich der Verlag das Recht der Vervielfältigung, Verbreitung und Übersetzung innerhalb der gesetzlichen Schutzfrist vor. Auszugsweiser Nachdruck oder photomechanische u. a. Wiedergabe sind an eine schriftliche Genehmigung des Verlages gebunden. Manuskripte, Zeichnungen und Photos von veröffentlichten Arbeiten werden nur auf den besonderen und bis zum Zeitpunkt der Veröffentlichung mitgeteilten Wunsch des Autors zurückgesandt.

Bei der unverlangten Einreichung von Manuskripten wird empfohlen, eine Befürwortung eines der Herausgeber oder Redaktionsausschußmitglieder beizufügen.

Manuskripte sind einzureichen bei der Redaktion, 1086 Berlin, Leipziger Str. 3–4.

2. Form der Manuskripte

Die Autoren werden gebeten:

2 Manuskripte, eines davon in Original-Maschinenschrift, einseitig und mit 2zeiligem Abstand einzureichen; gut leserliche Photokopien sind zulässig; auf dem 1. Blatt (unter Weglassung der Bezeichnung)

den Titel der Arbeit
Vorname-Initialen und Namen sämtlicher Autoren
gegebenenfalls die Institution, in der sie tätig sind
eine kurze Zusammenfassung der Arbeit von höchstens 15 Zeilen
Keywords in alphabetischer Reihenfolge und
als Fußnote die Anschrift der Institution anzugeben;

wenn möglich, auf gesonderten Blättern Namen, Titel der Arbeit und Zusammenfassung in den beiden anderen obengenannten Sprachen 2fach beizufügen;

Literaturhinweise (fortlaufend numeriert), Fußnoten, Tabellen, Abbildungen (in 2–3facher Veröffentlichungsgröße), Abbildungsunterschriften (getrennt von den Abbildungen mit Zeichenerklärungen, Quellenangaben, falls erforderlich, und gegebenenfalls mit einem Vermerk für besondere Wünsche bezüglich der Abbildungsgröße), private und Dienstanschrift der Verfasser mit akademischen Titeln und ausgeschriebenen Vornamen auf besonderen Blättern ans Ende des Manuskriptes zu stellen;

Wünsche für den Kolumnentitel (Seitenüberschrift) bekanntzugeben;

Absätze durch Einzüge von mindestens 3 Anschlägen kenntlich zu machen; die Maßeinheiten des SI-Systems („Systeme International") zu verwenden; Korrekturen im Manuskript unmittelbar an der zu korrigierenden Stelle anzubringen;

Formeln deutlich lesbar und möglichst groß einzutragen;

Farbabbildungen und nachträgliche Änderungen oder Ergänzungen in Abbildungen zu vermeiden;

hervorzuhebende Textstellen zu unterstreichen, und zwar halbfetten Satz mit einer geraden schwarzen Linie, kursiven Satz (Schrägschrift) mit einer blauen Wellenlinie (alle Eigennamen). Sperrungen gestrichelt;

besondere Buchstaben durch farbige Unterstreichung zu kennzeichnen, und zwar Formelzeichen in

Griechisch-Kursiv (α, β, ...) rot,
Kursiv (A, B, ...) blau
halbfetter Kursiv ($\boldsymbol{A}$, $\boldsymbol{B}$, ...) 2 Linien blau (Vektoren),
halbfetter Grotesk (**A**, **B**, ...) 2 Linien violett (Tensoren);

bei Literaturangaben auf Vollständigkeit zu achten. In nachstehender Reihenfolge müssen angegeben sein:

a) bei *Zeitschriften*: Vorname-Initialen und Namen sämtlicher Autoren, Titel der Zeitschrift (nach „Periodica Chimica" oder möglichst wenig gekürzt), Bandnummer (unterstrichen), Jahreszahl (in Klammern) Heftnummer (nur bei Zeitschriften, die jedes Heft mit S. 1 beginnen) sowie die Seitenzahl, auf der die zitierte Arbeit beginnt.

b) bei *Büchern*: Initialen der Vornamen und Name des Autors (Herausgeber werden durch den Zusatz „ed." gekennzeichnet), voll ausgeschriebener Tite des Buches in der Originalsprache, evtl. Bandnummer, voll ausgeschriebener Name des Verlages, Nummer der Auflage (außer der ersten), Erscheinungsort(e), Erscheinungsjahr (hier nicht in Klammern setzen!) und evtl. Kapitelnummer, Seitenzahl u. ä.

c) bei *Sammelbänden, Tagungsberichten*: Es soll zusätzlich der Titel des zitierten Beitrags angegeben werden. Die Seitenzahl muß unbedingt angegeben werden. Aus dem Zitat muß hervorgehen, ob es sich um einen Tagungssammelband oder um einen noch nicht veröffentlichten Vortrag handelt. Zwischen Tagungsort und Erscheinungsort des entsprechenden Tagungsberichts ist zu unterscheiden.

3. Korrekturen

Der Autor erhält zwei Umbruchabzüge, von denen einer möglichst schnell korrigiert an die Redaktion zurückgereicht werden soll. Auf Wunsch des Autors kann die Korrektur auch von der Redaktion gelesen werden.

Durch nachträgliche Manuskriptänderungen verursachte Korrekturkosten gehen, sofern sie 10% der Satzkosten übersteigen, zu Lasten des Autors.

Der Autor erhält 50 Sonderdrucke kostenlos. Weitere Sonderdrucke können bis zur Rückgabe der Korrekturabzüge auf Kosten des Autors bei der Redaktion bestellt werden.

Bezugsmöglichkeiten

Bestellungen sind zu richten

- **in der DDR** an den Postzeitungsvertrieb unter Angabe der Kundennummer des Bestellenden oder an den AKADEMIE-VERLAG BERLIN, DDR-1086 Berlin, Leipziger Straße 3–4, PF-Nr. 1233;
- **im sozialistischen Ausland** an eine Buchhandlung für fremdsprachige Literatur oder an den zuständigen Postzeitungsvertrieb;
- **in der BRD und Berlin(West)** an eine Buchhandlung oder an die Auslieferungsstelle KUNST UND WISSEN, Erich Bieber OHG, Wilhelmstraße 4–6, D-7000 Stuttgart 1;
- **in den übrigen westeuropäischen Ländern** an eine Buchhandlung oder an die Auslieferungsstelle KUNST UND WISSEN, Erich Bieber GmbH, Dufourstraße 51, CH-8008 Zürich;
- **im übrigen Ausland** an den Internationalen Buch- und Zeitschriftenhandel; den Buchexport Volkseigener Außenhandelsbetrieb der Deutschen Demokratischen Republik, DDR-7010 Leipzig, Postfach 160; oder an den AKADEMIE-VERLAG BERLIN, DDR-1086 Berlin, Leipziger Straße 3–4.

Zeitschrift „Isotopenpraxis"

Herausgeber: Zentralinstitut für Isotopen- und Strahlenforschung der Akademie der Wissenschaften der DDR.
Verlag: Akademie-Verlag Berlin, DDR-1086 Berlin, Leipziger Straße 3–4, PF-Nr. 1233; Fernruf: 2236221 oder 2236229; Telex-Nr.: 114420;
Bank: Staatsbank der DDR, Berlin, Kto.-Nr.: 6836-26-20712.
Chefredakteur: Prof. Dr. H. Hübner, Leipzig.
Anschrift der Redaktion der Zeitschrift Isotopenpraxis: Prof. Dr. H. Hübner, DDR-7010 Leipzig, Karl-Marx-Platz, Hauptpostamt, PSF 370; Fernruf: 281018.
Veröffentlicht unter der Lizenznummer 1527 des Presseamtes beim Vorsitzenden des Ministerrates der Deutschen Demokratischen Republik.
Gesamtherstellung: VEB Druckerei „Thomas Müntzer", DDR-5820 Bad Langensalza.
Erscheinungsweise: Die Zeitschrift „Isotopenpraxis" erscheint monatlich. Die 12 Hefte eines Jahrganges bilden einen Band. Bezugspreis je Band 240,— DM zuzüglich Versandspesen; Preis je Heft 20,— DM. Der gültige Jahresbezugspreis für die DDR ist der Postzeitungsliste zu entnehmen.
Bestellnummer dieses Heftes: 1079/21/8.

AN (EDV) 12831
00900

Isotopenpraxis 21 (1985) 8, S. 261 bis 268

Normal Coordinate Treatment of Liquid Water and Calculation of Vapor Pressure Isotope Effects

B. Gellai

(Central Research Institute for Physics)[1])

W. A. Van Hook

(Chemistry Department, University of Tennessee)[2])

A vibrational analysis of liquid water is reported, assuming a completely hydrogen-bonded network with continuously varying strengths of the hydrogen bonds. Frequency distribution calculations are made for intramolecular stretching and bending modes and for the intermolecular frequency region. The calculated distributions are compared with the experimental spectroscopic ones. As another test vapor pressure isotope effects are calculated from the theoretical distributions for some isotopic water molecules. Results are compared with ones of other authors obtained from a mixture model.

Es wird über die Schwingungsanalyse von Wasser und einigen Isotopen (T, D, ^{17}O, ^{18}O) berichtet. Es wurde ein vollständig wasserstoffverbundenes Netzwerk mit veränderlicher Bindungskraft angenommen. Die Häufigkeitsverteilung wurde sowohl für den innermolekularen (Valenz- und Deformationsschwingungsfrequenz) als auch für den intermolekularen Frequenzbereich (Schiebungs- und Schwingungsfrequenz) berechnet und mit den experimentellen specktroskopischen Häufigkeitsverteilungen verglichen. Aus den theoretischen Häufigkeitsverteilungen wurde der Dampfdruck-Isotopeneffekt für einige isotope Wassermoleküle berechnet. Die Ergebnisse wurden mit denen anderer Autoren für ein auf die Struktur des Wassers bezogenes Modell (ein Mischungsmodell) verglichen.

Keywords

dynamic function studies; isotope effects; normal-mode analysis; partition function; thermal equilibrium; vapor pressure; water vapor

1. Introduction

In spite of the relentless efforts of scientists since the days of *Roentgen* [1], there is still no generally accepted theory of the structure of liquid water, to say nothing of a general theory of the liquid state. In its absence various models have been proposed that can be classified under two broad headings: viz. mixture models, continuum models.

Mixture models [2–4] describe liquid water as an equilibrium mixture of several molecular species with a different number of hydrogen bonds per molecule, whereas continuum models [5–10] consider liquid water to be a completely hydrogen-bonded network. If a vibrational analysis of liquid water is performed assuming one of the above-mentioned models the frequency distribution of the intra- and intermolecular vibrations can be calculated. Comparison between the theoretical and the experimental spectroscopic distributions can give information on the validity of the assumed model. In addition to the spectroscopic comparison the calculation of certain thermodynamic properties which depend on the frequencies of liquid water, together with a comparison with experimental data can also be useful. One such thermodynamic property is the vapor pressure ratio of isotopically substituted molecules. This ratio is often discussed in the framework of the oversimplified cell modell [11], using the harmonic approximation. The condensed phase is modelled in terms of a single and assumed average molecule which executes its motions in a harmonic and isotropic cell imposed by the intermolecular field of its neighbors [12]:

$$\frac{P'}{P} = \prod_{\substack{\text{internal}\\ \text{frequencies}}}^{3n-6} \left[\frac{(u_i/u_i')_c}{(u_i/u_i')_g}\right] \left|\frac{\exp(u_i' - u_i)_{c/2}}{\exp(u_i' - u_i)_{g/2}}\right] \times \left[\frac{(1-\exp(-u_i')_c)/(1-\exp(-u_i)_c)}{(1-\exp(-u_i')_g)/(1-\exp(-u_i)_g)}\right] \times \prod_{\substack{\text{external}\\ \text{condensed}\\ \text{frequencies}}}^{6} \frac{u}{u'}\left[\exp\left(\frac{u_i' - u_i}{2}\right)\right] \times \left[\frac{1-\exp(-u_i')}{1-\exp(-u_i)}\right] \tag{1}$$

where $u_i = \frac{h\nu_i}{kT}$, ν_i is the i-th normal frequency and n is the number of atoms per molecule. Using Eqn. (1), vapor pressure isotope effects prove to be a very sensitive probe for measuring the net frequency shift on condensation. For example, consider OH substitution at 300 K where one readily calculates that a 1 cm^{-1} change in the shift of an OH stretching motion is equivalent to $\Delta \ln (P_H/P_D) = \pm 2.4 \times 10^{-3}$.

In the past ten years there have been several computations of general spectroscopic features for water using both mixture [13–16] and continuum models [17–20]. Using the continuum model *Curnutte* and *Bandekar* calculated the intramolecular frequency distributions of the molecule HDO as well as intermolecular frequency distributions for $H_2^{16}O$ [19–20]. *O'Ferrall* et al. [16] calculated liquid phase intra- and intermolecular frequencies and

Addresses: [1]) H-1525 Budapest, P.O. Box 49, Hungary; [2]) Knoxville, TN 37996 USA

vapor pressures for the molecules $H_2^{16}O$ and $D_2^{16}O$ using a nine-atom mixture model.

In an earlier paper [21] we reported a Monte Carlo calculation of the vapor pressure isotope effects (VPIE) for some isotopic water molecules. In this paper we describe the method as it has been applied to liquid water in detail. The calculated frequency distributions are critically compared with the experimental spectroscopic ones and we make a comparison between our VPIE results and ones of other authors [16].

2. Normal coordinate treatment of liquid water

For computing economy and mathematical simplicity, the inter- and intramolecular portions of the problem were solved independently using the appropriate continuum models [19, 20].

2.1. *Computation of intramolecular frequency distribution*

We used the following symmetry coordinates [22]:

$$\left.\begin{aligned} u_1 &= x'_1 - [(m_2x'_2 + m_3x'_3)/(m_2 + m_3)] \\ u_2 &= y'_1 - [(m_2y'_2 + m_3y'_3)/(m_2 + m_3)] \\ u_3 &= x'_2 - x'_3 \end{aligned}\right\} \qquad (2)$$

where the primed quantities are rectangular components of displacement from the equilibrium position and $Z'_i = 0$. Identification of the internal coordinates of a general bent XYZ molecule is given in Fig. 1. For the molecules $H_2^{16}O$ and $D_2^{16}O$ Eqns. (2) simplify appropriately.

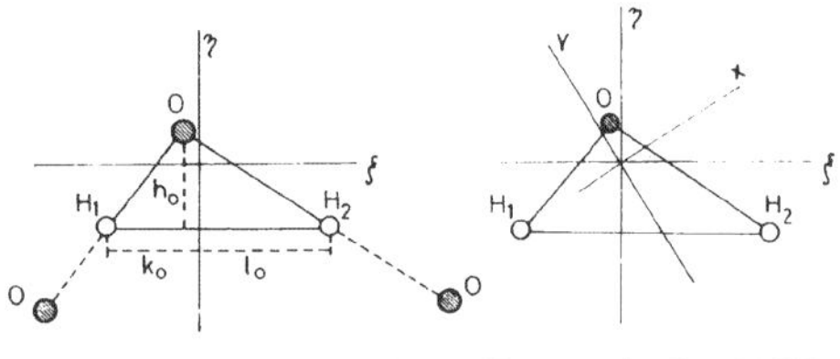

Fig. 1. Identification of internal coordinates of a bent XYZ molecule [22]

The kinetic and potential energy in terms of the symmetry coordinates (2) follow:

$$\left.\begin{aligned} T &= (1/2)\,[\mu_{11}\dot{u}_1^2 + \mu_{22}\dot{u}_2^2 + \mu_{33}\dot{u}_3^2 + 2\mu_{13}\dot{u}_1\dot{u}_3 \\ &\quad + 2\mu_{12}\dot{u}_1\dot{u}_2 + 2\mu_{23}\dot{u}_2\dot{u}_3] \\ V &= (1/2)\,[f_{11}u_1^2 + f_{22}u_2^2 + f_{33}u_3^2 + 2f_{12}u_1u_2 \\ &\quad + 2f_{13}u_1u_3 + 2f_{23}u_2u_3] \end{aligned}\right\} \qquad (3)$$

Coefficients μ_{ij} ($i_{j} = 1, 2, 3$) are the elements of the kinetic energy matrix and can be calculated from the mass and geometric parameters of the molecule. The numerical values of these parameters are given in Tab. 1. The quantities f_{ij} ($i, j = 1, 2, 3$) are generalized force constants defined by Eqn. (4).

$$f_{ij} = [(A_iA_j) + (B_iB_j)]\,f_r + C_iC_jf_\beta + [(A_i + B_i)\,C_j + (A_j + B_j)\,C_i]\,f_{r\beta} + (A_iB_j + A_jB_i)\,f_{rr} \qquad (4)$$

The coefficients A_i, B_i, C_i are the elements of the matrix of the transformation between the internal and the symmetry coordinates. The diagonal force constants f_r and f_β are functions of the oxygen-oxygen distance. The formula for f_r can be derived by using the linear model of the *Lippincott-Schroeder* potential function $V = V_1 + V_2$ [23], where

$$\begin{aligned} V_1 &= D[1 - \exp(-n(r - r_0)^2/2r)] \qquad (5) \\ V_2 &= D^*\{1 - \exp[-n^*(R - r - r_0^*)^2/2(R - r)]\} \end{aligned}$$

$D = 3.4328 \times 10^{-11}$ kJ/mol is the bond dissociation energy, the parameters $n = 9.18 \times 10^8\ \text{cm}^{-1}$ and $n^* = 13.32\ \text{cm}^{-1}$ are related to the ionization potentials of the atoms making up the bond and the value of D^* can be obtained from the relation $n^*D^* = nD$. $r_0^* = r_0 = 0.097$ nm is the equilibrium O—H distance, r and R are a given O—H and a given 0 ... 0 distance respectively.

From the conditions

$$\left(\frac{\partial V}{\partial r}\right)_{eq} = 0\,; \qquad \frac{\partial^2 V}{\partial r^2} = f \qquad (6)$$

characteristic of the motion of the hydrogen atom along the axis of the hydrogen bond we obtain for the intramolecular stretching force constant

$$\begin{aligned} f_r = {} & \left(\frac{nD}{r^3}\right) \times \left[\exp\left(\frac{-n\,\Delta r^2}{2r}\right)\right] \times \left[r_0^2 - \frac{n\,\Delta r^2}{4r}(r + r_0)^2\right] \\ & + \left[\frac{n^*D^*}{(R - r)^3}\right] \times \left[\exp\left(-\frac{n^*(R - r - r_0^*)^2}{2(R - r)}\right)\right] \\ & \times \left[r_0^{*2} - n^*(R - r - r_0^*)^2\,\frac{(R - r + r_0^*)^2}{4(R - r)}\right] \end{aligned} \qquad (7)$$

Because we were not able to obtain an expression relating the 0—0 distance to the bending force constant, the following empirical relation was substituted:

$$F_\beta = f_\beta^\circ \exp\{a[(R_0/R) - 1]\} \qquad (8)$$

Here $R_0 = 0.285$ nm is the equilibrium 0—0 distance, R is a given 0 ... 0 distance and $f_\beta^\circ = 2.019\ \text{N cm}^{-1}$ is the bending force constant at equilibrium. The coefficient $a = 0.3465$ was chosen to reproduce the experimental bandwith and maximum (298 K) of the bending mode of H_2O. After marking off the range of the accessible 0 ... 0 distance $0.318\ \text{nm} \geqq R_i \geqq 0.256$ nm (the upper bound is arbitrary), several thousand molecular configurations were selected by picking 0 ... 0 distance pairs from the allowable range for the 0 ... 0 distances according to the

Tab. 1. Masses and moments of inertia of some isotopic water molecules

Atom	Atomic masses (AWU) [25]	Molecule	Moments for inertia in equilibrium geometry, ($kg\,m^2$) I_{xx}	I_{yy}	I_{zz}
H	1·00783	$H_2^{16}O$	2.18671×10^{-47}	3.15787×10^{-47}	0.971157×10^{-47}
$^2H(D)$	2·01410	$D_2^{16}O$	4.37006×10^{-47}	6.11581×10^{-47}	1.74575×10^{-47}
$^3H(T)$	3·01605	$T_2^{16}O$	6.54402×10^{-47}	8.92040×10^{-47}	2.37637×10^{-47}
^{16}O	15·99491	$H_2^{17}O$	2.18671×10^{-47}	3.17012×10^{-47}	0.983411×10^{-47}
^{17}O	16·99914	$H_2^{18}O$	2.18638×10^{-47}	3.16367×10^{-47}	0.97729×10^{-47}
^{18}O	17·99916				
	HOH ∢ = 109° 28′		$R_0 = 0.285$ nm		$r_0 = 0.099$ nm

method of *Metropolis* et al. [24]. The main diagonal stretching and bending constants were changed according to randomly picked 0 ... 0 distances using Eqns. (7) and (8); the numerical value of the off-diagonal force constants was chosen to reproduce the water vapor fundamental frequencies with $R = \infty$ and can be seen in Tab. 2. The intramolecular frequencies of a given molecule can be obtained as the solution of the following eigenvalue problem:

$$\begin{bmatrix} \mu_{11}\lambda - f_{11} & \mu_{12}\lambda - f_{12} & \mu_{13}\lambda - f_{13} \\ \text{symm.} & \mu_{22}\lambda - f_{22} & \mu_{23}\lambda - f_{23} \\ & & \mu_{33}\lambda - f_{33} \end{bmatrix} \tag{9}$$

where $\lambda_k = 4\pi^2\nu_k^2$ and ν_k ($k = 1, 2, 3$) is the k-th normal frequency of the molecule. The ordinate values of the observed 0 ... 0 pair correlation function [26–28] (Fig. 2) represent the probability that an O–H ... 0 bond has the corresponding 0 ... 0 length, so the probability of a random water molecule configuration is the product of the ordinates corresponding to the randomly chosen 0 ... 0 distances. These probability values were stored in subintervals or "boxes" having an appropriate length in a given range of frequencies. The stored probability values are defined as

$$a_i(\nu) = \text{Prob}\,[\nu_{i-1} < \nu \leq \nu_i] = \frac{N_i}{N} \tag{10}$$

where ν is a given frequency in a given frequency distribution; N_i is the number of ν such that $\nu_{i-1} < \nu \leq \nu_i$; and N is the total number of frequencies in the given frequency range [29]. Vibrational distributions are converted

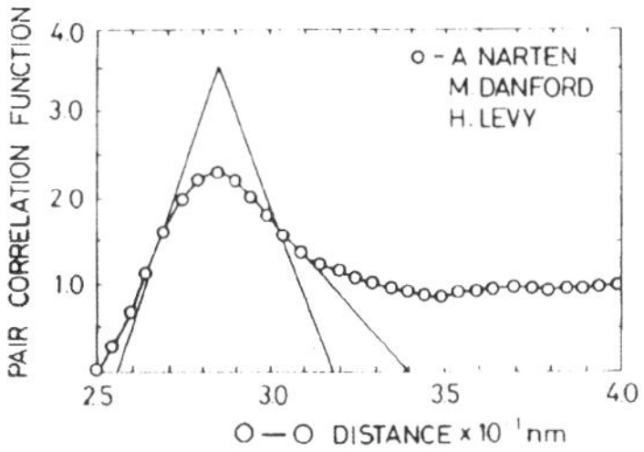

Fig. 2. The pair correlation function for liquid water at 25 °C vs. the oxygen-oxygen distance is denoted by circles [26]

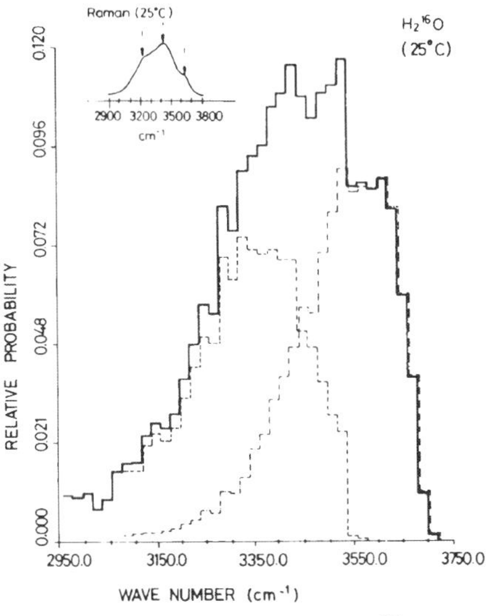

Fig. 3. Calculated frequency distribution of $H_2^{16}O$ in the OH stretching region

ν_1 (– – –) is the left; ν_3 (. – . – .) is to the right; $\nu_1 + \nu_3$ = solid line

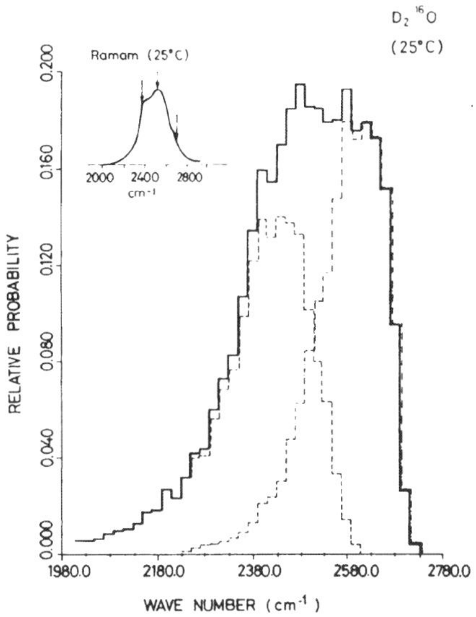

Fig. 4. Calculated frequency distribution of $D_2^{16}O$ in the OD stretching region

ν_1 (– – –) is the left; ν_3 (. – . – .) is to the right; $\nu_1 + \nu_3$ = solid line

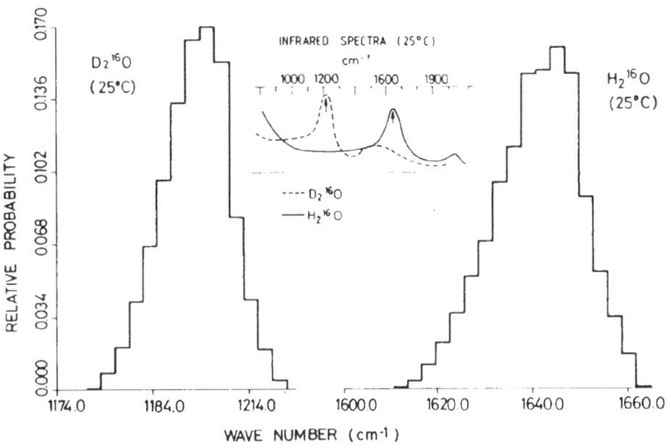

Fig. 5. Calculated frequency distribution in the bending region of $D_2^{16}O$ (to the left) and $H_2^{16}O$ (to the right)

into spectra by taking the intensity of the band at each frequency as being determined by the number of oscillators having that particular frequency at any instant. The intramolecular stretching frequency distributions for $H_2^{16}O$ and $D_2^{16}O$ are presented in Fig. 3 and 4 respectively; Fig. 5 reviews the calculated spectra in the bending region. The histograms are drawn over 20 cm⁻¹ intervals for the stretching region and over 3 cm⁻¹ intervals for the bending region. The observed spectra are shown as insets in the figures. It can be seen that the calculated shapes and bandwidths ($\nu_1 + \nu_3$) correlate nicely with the experimental spectroscopic distribution. Assignments of the band maxima and a comparison with observed spectra and the results of *O'Ferrall* et al. [16] will be provided in the last section of the paper.

2.2. *Computation of intermolecular frequency distribution*

The local structure model for our computation was the same as that used by *Bandekar* and *Curnutte* [20]. It consists of a central rigid water molecule tetrahedrally coordinated to four nearest neighbours surrounded by a rigid cage of next nearest neighbors (Fig. 6). The influence

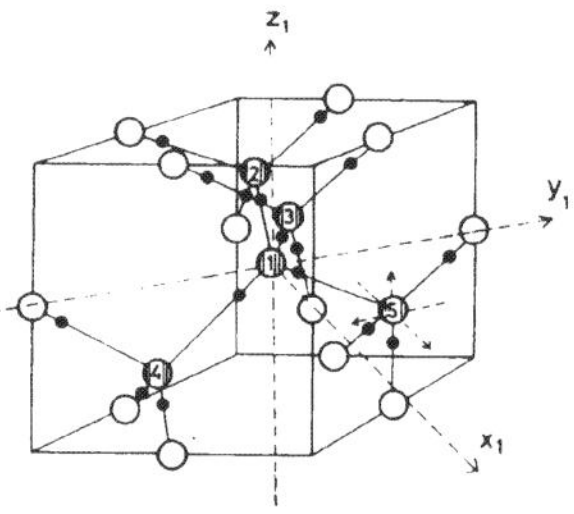

Fig. 6. Local structure unit $(H_2O)_5$ used for computing the intermolecular frequency distribution of liquid water

of these last neighbors is accounted for by the average force they exert on the (five molecule) local structure unit. The expression for the kinetic energy of the unit is

$$2T = M \sum_{i=1}^{5} (\dot{x}_i^2 + \dot{y}_i^2 + \dot{z}_i^2) + \sum_{j=1}^{5} (I_{xx} \dot{R}_{xj}^2 + I_{yy} \dot{R}_{zj}^2 + I_{zz} \dot{R}_{yj}^2 - 2 I_{yz} \dot{R}_{yj} \dot{R}_{zj}) \quad (11)$$

where R_{xj}, R_{yj} and R_{zj} denote rotational displacements (radians) about the x, y and z axes for molecule j; the quantities x_i, y_i, z_i are translational displacements from equilibrium. I_{xx}, I_{yy} and I_{zz} are the principal moments of inertia. Their numerical values in the equilibrium geometry can be seen in Tab. 1.

The valence-bond form of the potential energy of the unit in the harmonic approximation is

$$2V = \sum_{i=1}^{16} (f_{\rho i} \varrho_i^2 + f_{\sigma i} \sigma_i^2 + f_{\tau i} \tau_i^2) + \sum_{j=1}^{5} (pE)_j (R_{xj}^2 + R_{yj}^2) \quad (12)$$

where ϱ, σ and τ represent the stretching, in-plane bending, and out-of-plane bending displacement of hydrogen bonds from their position of equilibrium (Fig. 7).

Tab. 2. Force field and dipole moment for liquid water in equilibrium geometry (25 °C)

Force constants	Dipole moment
Stretching force constant:	Equilibrium
$f_\rho = 0{\cdot}192 \times 10^2$ Nm^{-1}	
"In-plane"bending:	(pE) value:
$f_\sigma / r_0 R_0 = G = 0{\cdot}037 \times 10^{-8}$ Nrd^{-1}	
"Out-of-plane" bending:	$(pE)_0 = 0{\cdot}119 \times 10^{-17}$ J
$f_\tau / r_0 R_0 = H = 0{\cdot}03675 \times 10^{-1}$ Nrd^{-1}	
Bandekar-Curnutte [20]	
f_ρ = (the above value)	
$G = 0{\cdot}024 \times 10^{-8}$ Nrd^{-1}	
$H = 0{\cdot}03275 \times 10^{-8}$ Nrd^{-1}	

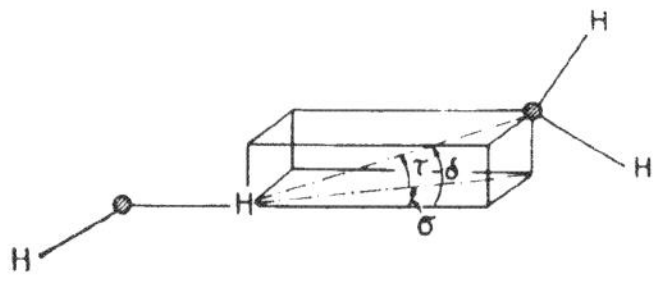

Fig. 7. Hydrogen bond coordinates
The Fig. was taken from [20].

Algebraic expressions for f_ρ, f_σ and f_τ, the corresponding force constants, were obtained from the *Lippincott-Schroeder* potential function which is valid for a bent hydrogen bond [30]; they were scaled by multiplication with a constant to yield the numberical value given in Tab. 2. These values were chosen as a means of producing the convergence of the calculated VPIE (see next section). The last term in Eqn. (12) is the interaction energy of the electric dipole moment, p, of a molecule in the liquid state with the local electric field, E, due to its neighbors. The product pE was varied according to the average value of the reciprocal of the 0 ... 0 distance cubed, where the average is taken over the four hydrogen bonds of the molecule of interest. The potential energy expression was transformed from valence bond coordinates to cartesian displacement coordinates using the geometrical relationships between the two sets of coordinates [31], and *Euler-Lagrange* equations were constructed for the local structure unit $(H_2O)_5$. Frequencies obtained as the solution of a 30 × 30 eigenvalue problem first constructed for the symmetric configuration (Fig. 6) with all 0 ... 0 distances set to 0·285 nm. The resulting frequencies for this symmetric configuration are compared with the assignments from observed spectra [32] and the frequencies calculated by *Bandekar* and *Curnutte* [20], for $H_2^{16}O$ in Tab. 3, and for $D_2^{16}O$ in Tab. 4. The intermolecular frequency distributions of water in the liquid phase as obtained by various experimental techniques can be seen in Fig. 8. The calculated frequency distribution of 250 random $(H_2^{16}O)_5$ configurations obtained by assigning the 16 hydrogen bonds randomly for each configuration was obtained by accumulating the probability just as was done in the case of intramolecular vibrations. It can be seen for $H_2^{16}O$ in Fig. 9 and for $D_2^{16}O$ in Fig. 10. It is clear from the figure that it is almost impossible to identify band maxima with separate librational or hindered translational classes. This raises no difficulty since the present method for computing the VPIE does not need such specific assignment in this region.

Tab. 3. Calculated intermolecular frequencies for the symmetric unit $(H_2^{16}O)_5$ with all 0 ... 0 distances equal to 0·285 nm are compared with the experimental findings (25 °C; cm^{-1})

Raman and hyper Raman [32]	Infrared [32]	Present calculation	*Bandekar-Curnutte* [20]
		Frequencies of hindered translations	
~ 60	~ 70	70, 73, 75	71, 77, 75
166	170	160, 162, 162, 163, 163, 163, 164, 164, 164	164, 164, 160, 163, 164, 162, 165, 165, 165
		217, 218, 218	217, 219, 219
		Frequencies of hindered rotations	
425 – 450		394, 394, 394, 394, 394	450, 450, 450, 452, 452
550		577, 577, 577, 578, 578	553, 553, 554, 554, 553
	685	611, 611, 611, 612, 612	
720 – 740			722, 722, 722, 723, 723

Tab. 4. Calculated intermolecular frequencies for the symmetric unit ($D_2^{16}O$) with all O ... O distances equal to 0·285 nm are compared with the experimental findings (25 °C; cm⁻¹)

Raman and hyper Raman [32]	Infrared [32]	Present calculation	*Bandekar-Curnutte* [20]
		Frequencies of hindered translations	
~ 60		66, 69, 71	68, 74, 72
~ 175	165	152, 153, 154, 155, 155, 155, 155, 155, 155	157, 157, 152, 155, 157, 154, 157, 157, 157
		205, 206, 207	206, 208, 207
		Frequencies of hindered rotations	
305, 350		282, 282, 282, 283, 284	319, 320, 318, 320, 319
415		427, 427, 431, 431, 431	397, 399, 397, 398, 397
		432, 432, 432, 436, 437	
500, 570	505		538, 539, 538, 539, 538

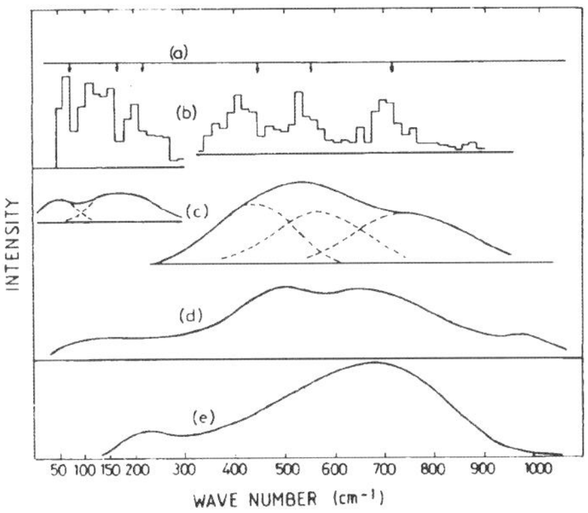

Fig. 8. Experimental intermolecular frequency distributions in water (a) Raman scattering at 25 °C [32, 33]; (b) Inelastic neutron scattering at 25 °C [34]; (c) Far infrared results at 25 °C [35]
The Fig. was taken from [20].

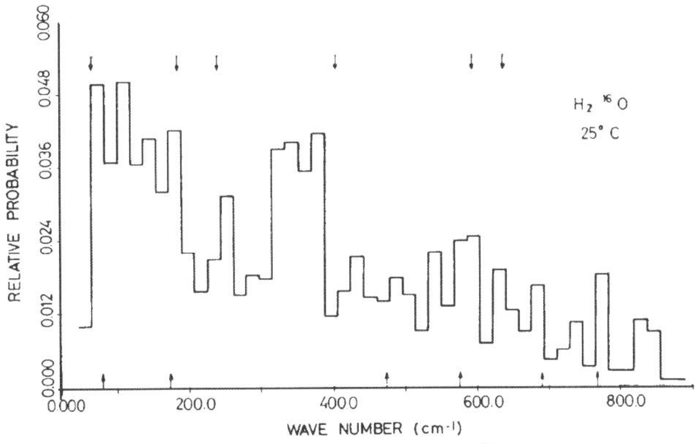

Fig. 9. Calculated frequency distribution of $H_2^{16}O$ in the intermolecular region

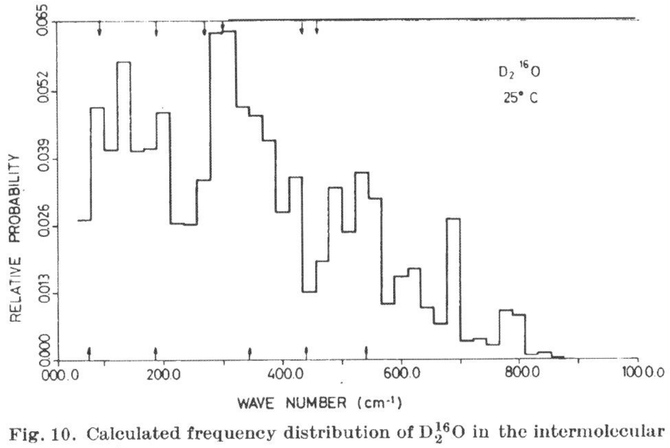

Fig. 10. Calculated frequency distribution of $D_2^{16}O$ in the intermolecular region

3. Computation of the vapor pressure isotope effect

Taking the logarithm of Eq. (1) and using the stored probabilities as calculated in the previous section we obtain

$$\ln \frac{P'}{P} = \sum_{j=1}^{3n-6} \left\{ \sum_{i=1}^{n_j} a_{ji} \left[\ln \frac{u_{ji} \exp(-u_{ji}/2)}{1 - \exp(-u_{ji})} \right] - \sum_{i=1}^{n_j'} a_{ji}' \left[\ln \frac{u_{ji}' \exp(-u_{ji}'/2)}{1 - \exp(-u_{ji}')} \right] \right\} - \ln (s/s' f)_{gas}$$
$$+ 6 \left\{ \sum_{i=1}^{n} a_i \left[\ln \frac{u_i \exp(-u_{i/2})}{i - \exp(-u_i')} \right] - \sum_{i=1}^{n'} a_i' \ln \left[\frac{u_i' \exp(-u_i'/2)}{1 - \exp(-u_i')} \right] \right\} \tag{13}$$

where n_j and n represent the number of boxes in the range of the j-th intramolecular and intermolecular frequency distribution respectively. Primes denote values for the lighter isotopes. The probability values fulfil the following normalization requirements:

$$\sum_{i=1}^{n_j'} a_{ji}' = \sum_{i=1}^{n_j} a_{ji} = 1 \qquad j = 1, 2, \ldots, 3n - 6 \tag{14}$$

$$\sum_{i=1}^{n'} a_i' = \sum_{i=1}^{n} a_i = 1$$

In Eqn. (13) the term $\ln (s/s'f)_{gas}$ contains the frequencies of the gas phase. The main diagonal elements of the force constant matrix, F_{gas}, necessary for calculating these frequencies were obtained by substituting the maximum allowable O ... O distance, $R_{max} = 0·318$ nm into the L.S. potential function (5) and into Eqn. (8). The off diagonal elements of the matrix were taken from *Curnutte* and *Bandekar* (Tab. 2). It was assumed for such separated water molecules the intermolecular forces can be neglected. Thus the model is an approximation of the gas phase and it is more appropriate to label the term $\ln (s/s'f)_{gas}$ "isolated" instead of "gas". The frequencies of some "isolated" isotopic water molecules are compared with the real gas phase frequencies in Tab. 5. The calculated frequencies differ from the exact measured gas phase frequencies but we emphasize that the point which interests us is the calculation of properly weighted gas-liquid frequency shifts, and not the duplication of the exact gas phase frequencies. In addition, it should be noted that the numerical value of R_{max} (0·318 nm in this paper) is arbitraty; if a larger value were to be chosen the gas phase frequencies might be better approximated.

Another important feature of the present calculation of the VPIE is that the whole intermolecular frequency

Tab. 5. Comparison of "isolated" and gas phase frequencies

Phase	$H_2^{16}O$ [36]	$D_2^{16}O$ [36]	$T_2^{16}O$ [37]	$H_2^{18}O$ [38]	$H_2^{17}O$
"Isolated"	3712·87	2720·38	2296·99	3697·74	3704·85
	3614·02	2605·26	2169·23	2606·39	3609·98
	1612·52	1180·16	992·74	1605·83	1608·98
Gas	3755·8	2788·1	2370·0	3741·6	–
	3656·7	2671·5	–	3649·7	–
	1594·6	1178·3	995·5	1588·2	–

region is treated as a group in order to avoid difficulties of assignment. This can be achieved by constructing the last term in Eqn. (13) in an appropriate manner.

In terms of the above approximations the computation of the VPIE was performed as follows. First ln $(s/s'f)_{gas}$ was calculated by using the frequencies of the "isolated" molecules (Tab. 5). Then, assuming that 20000 intramolecular configurations give reliable statistics, we computed the first term in Eqn. (13). Later the last ("intermolecular") term was calculated by generating 50 intermolecular configurations. We then tested the statistical convergence by increasing the number of configurations by units of 50, calculating the VPIE at each stage from Eqn. (13) with the a_{ji} and a_i values consistent with Eqn. (14), until we arrived at 250 configurations. Intermolecular bending force constants used by *Bandekar* and *Curnutte* were adjusted (see Tab. 2) to obtain convergence of the VPIE values by increasing the number of intermolecular configurations. The converged values are given in Tab. 6 and compared with experiment [39] and the earlier cell model calculations [40]. The convergence of the process

Tab. 6. Calculated and measured VPIE values. The values of ln (P_{H_2O}/P_X) are presentedin the Table at 25 °C

Molecule (X)	Cell model [40]	Present calc.	Experimental values with errors at 25 °C [39]
$D_2^{16}O$	0·1440	0·1410	0·1450 ± 2 × 10^{-4}
$T_2^{16}O$	0·1770	0·1750	0·1750 ± 5 × 10^{-4}
$H_2^{18}O$	0·0095	0·0098	0·0093 ± 6 × 10^{-4}
$H_2^{17}O$	0·0053	0·0054	0·0053 ± 4 × 10^{-4}

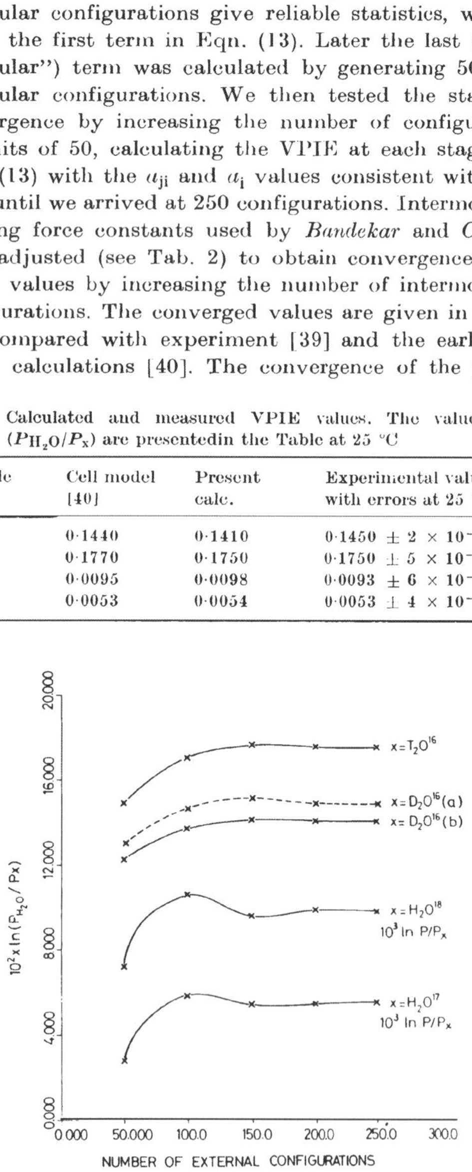

Fig. 11. Calculated VPIE values as a function of the number of intermolecular configurations (25 °C)

a – Intermolecular force constants: f_ρ = 0.192 × 10^2 Nm^{-1}, f_σ/r_0R_0 = 0.037 × 10^{-8} Nrd^{-1}, f_τ/r_0R_0 = 0.0367 × 10^{-8} Nrd^{-1} (see Tab. 2).

b – Intermolecular force constants: f_ρ = 0.211 × 10^2 Nm^{-1}, f_σ/r_0R_0 = 0·03816 × 10^{-8} Nrd^{-1}, f_τ/r_0R_0 = 0·03775 × 10^{-8} Nrd^{-1}

is presented graphically in Fig. 11. The convergence apparently good to ±0·002 unit at or above 150 configurations.

4. Interpretation and comparison with experiment and other results

In the experimental IR/Raman spectra of liquid water an intensity maximum occurs at 3410 ± 10^{-1}, an intense shoulder is found near 3225 ± 20 cm^{-1} and a weak shoulder appears near 3625 ± 20 cm^{-1} [41].T he shape of the joint distribution ($\nu_1 + \nu_3$) of the calculated bands (Fig. 3) reproduces these experimental features. In addition the feature arround 3550 ± 20 cm^{-1} correlates with an observed shoulder near 3565 cm^{-1} obtained from analysis of the argon-ion-laser excited Raman spectra of $H_2^{16}O$ [42]. The corresponding experimental features for liquid $D_2^{16}O$ occur near 2510 ± 10 cm^{-1}, 2390 ± 20 cm^{-1} and 2675 ± 20 cm^{-1} [41]. The band maxima of the sum of the individual symmetric ν_1 and asymmetric ν_3 bands (Fig. 4) correlate with these experimental findings. A fourth feature around 2580 ± 20 cm^{-1} correlates with the maxima 2590 cm^{-1} obtained by computer analysis of laser-Raman spectra in the OD stretching region from liquid $D_2^{16}O$ [41]. It should be emphasized that the above assignments are approximate. It is impossible to factor the experimental frequency distributions of HOH and DOD in the stretching region into individual ν_1 and ν_3 contributions. Fig. 5 reviews the calculated spectra in the bending region. Using Eqn. (8) the ν_2 band maxima are calculated and are found to be in agreement with experimental for $H_2^{16}O$ and $D_2^{16}O$, though the peak maximum for $D_2^{16}O$ is in less satisfactory agreement with the measurement [33]. This indicates a short-coming in Eqn. (8). Even so it is to be noted that the calculated VPIE's are not particularly sensitive to the value of the bending frequency.

In spite of the fact that the calculated spectra in the intermolecular frequency region show a very broad pattern with coarse features for $H_2^{16}O$ between 100 and 200 cm^{-1} and again around 300, 500 and 800 cm^{-1}, we have made no attempt to assign the frequencies to separate librational or hindered translational classes.

It is worthwhile to compare our results with those of *O'Ferrall* et al. [16]. They considered liquid water as a mixture of molecules with a different number of hydrogen bonds. A nine-atom model consisting of $H_2^{16}O$ surrounded tetrahedrally by two oxygens atom fragments and two OH groups is used for fully bound species and is called Model 1 (Fig. 12). The O–H covalent and non-bonded hydrogen bond lengths were taken as 1·0 and 1·76 Å, respectively, close to the values for ice. Molecules in lower states of hydrogen bonding are represented by removing one or both of the oxygens hydrogen-bonded to the hydrogens of the $H_2^{16}O$ and they are called Model 4 and Model 5 res-

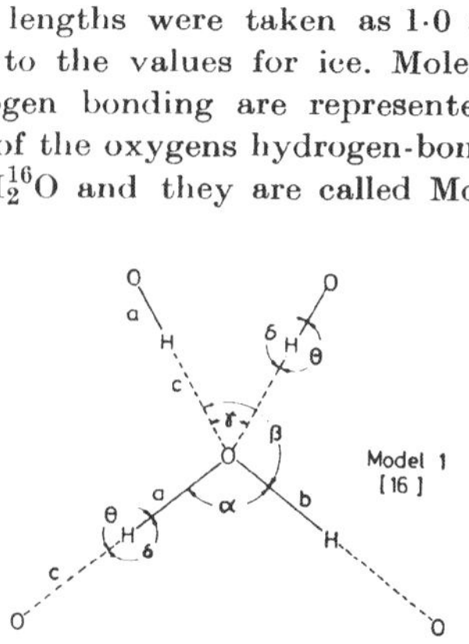

Fig. 12. Fully hydrogenbonded water
The Fig. was taken from [16].

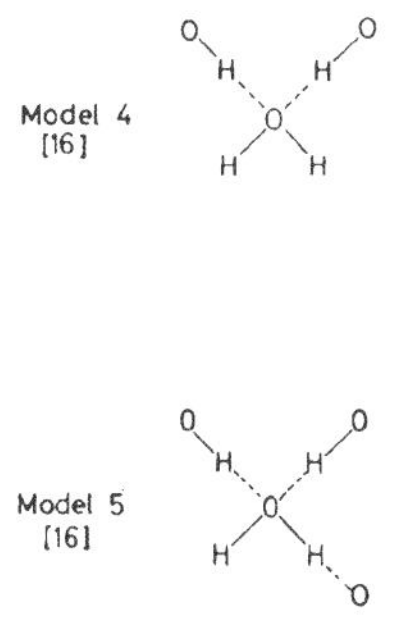

Fig. 13. Molecules in lower states of hydrogen bonding
The Fig. was taken from [16].

pectively (Fig. 13). The force constants of the Models are compared with ones used by us in Tab. 7 and the calculated internal and external frequencies for Models 1, 4 and 5 are presented in Tab. 8. The logarithms of the vapor pressure ratio, $P_{H_2^{16}O}/P_{D_2^{16}O}$, the cited authors obtained for Models 1 and 5 at 25 °C are $\ln (P_{H_2^{16}O}/P_{D_2^{16}O}) = 0{\cdot}241$ and $\ln (P_{H_2^{16}O}/P_{D_2^{16}O}) = 0{\cdot}176$ respectively; the experimental value is 0·142 [42, 43]. Tab. 6 clearly shows the consistency of our calculation as far as the effects in $D_2^{16}O$, $T_2^{16}O$, $H_2^{18}O$ and $H_2^{17}O$ are concerned. The calculated value of $\ln (P_{H_2^{16}O}/P_{D_2^{16}O})$ obtained by us is more reasonable than the one calculated by *O'Ferrall* et al. This can be attributed to certain advantages of the present continuum model. The first is the range of constants describing coupling to neighbors. This is explicitly included by making the force constant as a function of the nearest neighbor distances with the L.S. potential. Another advantage of the present cntinuum model is that the entire external frequency region can be treated as a whole, thus avoiding the separation of the librations which, as the above authors mention, "is largely artificial because there is a strong coupling between the modes" [16]. Finally it should be mentioned that in our method the computation of VPIEs was mainly achieved by adjusting the value of the in-plane and out-of-plane bending force constants. Apart from the relative contributions of the intramolecular frequencies, the calculated VPIEs paricularly depend on the librational modes. Their magnitude was determined principally by the in-plane and out-of-plane bending constants. The librational modes lie at the top of the intermolecular band and display more isotope dependence than do the hindered translations.

It would seem that the present continuum model is better suited to describing the main spectroscopic features of liquid water and to maintaining agreement between calculated and experimental VPIEs than the simple mixture model as used by *O'Ferrall* et al. It is of course vastly superior to the simple cell model [40] which makes no pretense of spectroscopic accuracy.

Tab. 7. Force constants used in this paper compared with those used by *O'Ferrall* et al. [16]

Force Constants used in this paper (R = 2·86 Å)	
f_r = 6·55 × 10^5 dynes/cm	Stretching: f_ρ = 1·92 × 10^4 dynes/cm
f_α = 0·2019 × 10^6 dynes/cm	"In-plane" bending: G = 3·700 × 10^3 dynes/cm
f_{rr} = −0·0794 × 10/ dynes/cm $f_{r\alpha}$ = 0·3877 → 10/ dynes/cm	"Out-of-plane" bending: H = 3·675 × 10^3 dynes/cm

Force constants used by *O'Ferrall* et al. [16]	
(The subcripts for the force constants follow the notation in Model 1)	
Model 1	$f_a = f_b$ = 6·55 × 10^5 dynes/cm f_c = 0·12 × 10^5 dynes/cm f_{ab} = −0·25 × 10^6 dynes/cm f_α = 0·65 × 10^5 dynes/cm f_β = 0·02163 × 10^5 dynes/cm f_γ = 0·09685 × 10^4 dynes/cm $f_\varphi = f_\delta$ = 0·02163 × 10^5 dynes/cm
Model 4	$f_a = f_b$ = 7·35 × 10^5 dynes/cm f_α = 0·73 × 10^5 dynes/cm The other force constants were as used in Model 1.
Model 5	f_a = 7·35 × 10^5 dynes/cm f_b = 6·55 × 10^5 dynes/cm f_α = 0·69 × 10^5 dynes/cm The other force constants were as used in Model 1.

Tab. 8. Calculated vibration frequencies for water with different number of hydrogen bonded hydroxyl group obtained by *O'Ferrall* et al. [16]

Frequencies cm⁻¹	Model 1		Model 4		Model 5	
	$H_2^{16}O$	$D_2^{16}O$	$H_2^{16}O$	$D_2^{16}O$	$H_2^{16}O$	$D_2^{16}O$
ν_1	3355	2425	3585	2590	3405	2470
ν_3	3550	2605	3670	2695	3675	2690
ν_2	1645	1200	1645	1205	1650	1200
ν_{lip}	715	530	410	307	710	515
	710	500	355	254	420	300
	645	470	230	173	380	280
ν_{def}	530	380	620	440	610	445
	530	380	600	435	600	435
	530	380	580	425	580	420
	(580)	(420)	580	420	530	380
ν_{trans}	168	161			162	155
	168	161	158	148	161	154
	167	161	137	132	126	122
	105	104				
ν_{bend}	54−35	53−55	41	41	59−39	58−38

Acknowledgements

Partial supports from the U.S. National Science Foundation and the Hungarian Institute of Cultural Relations as a part of the cooperative research program between the Central Research Institute for Physics, Budapest, and the University of Tennessee, Knoxville, are gratefully acknowledged. Generous grants of computer time from the Computing Centers of the University of Tennessee and the Central Research Institute for Physics are appreciated.

Received November 28, 1984

References

[1] *W. K. Roentgen*, Ann. Phys. Chim. **45** (1892) 92
[2] *G. Nemethy, H. A. Scheraga*, J. Chem. Phys. **36** (1962) 3382
[3] *A. Ben-Naim*, J. Phys. Chem. **69** (1965), 1922, 3240
[4] *R. P. Marchi, H. Eyring*, J. Chem. Phys. **38** (1964) 221
[5] *J. Lennard-Jones, J. A. Pople*, Proc. Roy. Soc. **A205** (1951) 155
[6] *J A. Pople*, Proc. Roy. Soc. **A205** (1951) 163
[7] *D. N. Glew*, J. Phys. Chem. **66** (1962) 605
[8] *M. Falk, T. A. Ford*, Can. J. Chem. **44** (1966) 1699
[9] *M. Falk, H. R. Wyss* J. Chem. Phys. **51** (1969) 5727
[10] *H. R. Wyss, M. Falk*, Can. J. Chem. **48** (1970) 607
[11] *J. Bigeleisen*, J. Chem. Phys. **34** (1961) 1435
[12] *M. J. Stern, W. A. Van Hook, M. Wolfsberg*, J. Chem. Phys. **39** (1963) 3179
[13] *H. S. Frank, W. Y. Wen*, Discussions, Faraday Soc. **24** (1957) 133
[14] *H. S. Frank*, Proc. Roy. Soc. London Ser. **A247** (1958) 481
[15] *F. Hajdu*, Acta Cryst. **A31** (1975) 157
[16] *R. A. More O'Ferrall, G. W. Koeppl, A. J. Kresge*, J. Amer. Chem Soc. **93** (1971) 1

[17] *J. B. Bryan*, "A Normal Coordinate Analysis of the Local Structure of Liquid Water for Interpretation of Far Infrared Spectra", Ph.D. Dissertation, Kansas State University, Manhattan 1969
[18] *J. A. Barker*, Ann. Rev. Phys. Chem. **14** (1965) 2145
[19] *B. Curnutte, J. Bandekar*, J. Mol. Spectrosc. **41** (1972) 500
[20] *J. Bandekar, B. Curnutte*, J. Mol. Spectrosc. **58** (1975) 169
[21] *B. Gellai, W. A. Van Hook*, Fluid Phase Equilibria **5** (1980) 19
[22] *W. H. Schaffer, R. P. Schumann*, J. Chem. Phys. **12** (1944) 504
[23] *E. R. Lippincott, R. Schroeder*, J. Chem. Phys. **23** (1955) 1099
[24] *N. Metropolis, A. W. Rosenbluth, M. H. Rosenbluth, A. H. Teller, E. Teller*, J. Chem. Phys. **21** (1953) 1087
[25] *G. S. Kell*, Water. A Comprehensive Treatise (Ed.: *F. Franks*) Plenum Press 1975, p. 377
[26] *A. H. Narten, M. D. Danford, H. H. Levey*, X-Ray Diffraction Data on Liquid Water in the Temperature Range 4 °C to 200 °C. Oak Ridge National Laboratory, Oak Ridge, TN, ORNL-3997, UC-4 (1966)
[27] *A. H. Narten, H. H. Levey*, J. Chem. Phys. **55** (1971) 2263
[28] *A. H. Narten*, J. Chem. Phys. **56** (1972) 5681
[29] *I. S. Bendat, A. G. Piersol*, Measurement and Analysis of Random Data. New York: John Wiley 1966, pp. 309–310
[30] *R. Schroeder, E. R. Lippincott*, J. Chem. Phys. **61** (1957) 921
[31] *J. N. Bandekar*, A Monte Carlo normal coordinate analysis treatment of intermolecular vibrations in liquid water. Ph.D. Dissertation, Kansas State University, Manhattan 1973
[32] *G. E. Walrafen*, in: "Hydrogen-bonded Solvent Systems" (Eds.: *A. K. Covington, P. Jones*) London: Taylor and Francis 1968, pp. 9–30
[33] *G. E. Walrafen*, J. Chem. Phys. **40** (1964) 3249; **47** (1967) 114
[34] *G. J. Safford, P. S. Leung, A. W. Naumann, P. C. Schaffer*, J. Chem. Phys. **50** (1969) 4444
[35] *C. Robertson, B. Curnutte, D. Williams*, Mol. Phys. **26** (1973) 183
[36] *T. Shimanouchi, I. Suzuki*, J. Chem. Phys. **42** (1964) 296
[37] *S. Pinchas, I. Laulicht*, Infrared Spectra of Labelled Compounds. London: Academic Press 1971, pp. 44–45
[38] *M. Majoube*, J. Chim. Phys. **68** (1971) 1423
[39] *G. Jancso, W. A. Van Hook*, Chem. Rev. **74** (1974) 689
[40] *W. A. Van Hook*, J. Phys. Chem. **72** (1968) 1234; J. Phys. Chem. **76** (1972) 3040
[41] *G. E. Walrafen*, Water. A comprehensive Treatise (Ed.: *F. Franks*) Plenum Press 1975, pp. 197–199
[42] *W. M. Jones*, J. Chem. Phys. **48** (1968) 207
[43] *L. Merlivat, R. Botter, G. Nief*, J. Chim. Phys. Physicochim. Biol. **60** (1963) 56; *F. D. Rossini, J. W. Knowlton, H. L. Johnston*, J. Res. Nat. Bur. Stand. **24** (1940) 369

Isotopenpraxis **21** (1985) 8, S. 268 bis 272

A Simple Model of the Global Distribution Process of Bomb-Produced Tritium*)

D. Hebert

(Bergakademie Freiberg, Sektion Physik)[1]

Als Maß des Abklingens des bis 1963 in die Stratosphäre eingetragenen Bomben-Tritiums wurde die zeitliche Abnahme der in Niederschlägen der Polargebiete beobachteten jährlichen Spitzenwerte der spezifischen Tritiumaktivität mit Hilfe eines Box-Modells der atmosphärischen Teilreservoire beschrieben. Die Transportkoeffizienten betragen für den interstratosphärischen Austausch 0,70 pro Jahr; für den Austausch zwischen Strato- u. Troposphäre 0.35 pro Jahr; für den Übergang Troposphäre-Ozean 4.15 pro Jahr. Der Tritiumgehalt der stratosphärischen Luftfeuchte klingt mit einer Halbwertszeit von 20 Monaten ab. Ein spezieller Antarktis-Kontinentaleffekt wird diskutiert.

The secular variation in the tritium content of precipitation, particularly in the polar regions, is investigated by mathematical treatment of a compartment model of exchangeable atmospheric reservoirs. Exchange coefficients are calculated to be 0.70 per year and 0.35 per year for interstratospheric exchange and exchange between stratosphere and troposhpere, respectively. The stratospheric half-live of bomb-produced tritium amounts to 20 months. A special Antarctic continental effect is discussed.

Keywords

antarctica; atmospheric precipitation; terrestrial atmosphere; global aspects; mathematical models; tritium

1. Problems

Until 1963 (the year in which the nuclear test can be treaty was concluded), a total of 816 kilogramms of tritium were injected into the stratosphere by thermonuclear tests [1].

This resulted in a substantial increase in the tritium content of precipitation in the northern hemisphere (Fig. 1) compared with the natural level of approximately 5 tritium units [2] (1 T.U. $\hat{=}$ 3·2 pCi/lH_2O $\hat{=}$ 0·12 Bq/kg H_2O).

The secular variation of the specific tritium activity of precipitation [3–6] is characterized by a rapid decrease after 1963, the rate of which dimished gradually; marked seasonal fluctuations with peak values in the spring; and a re-increase and subsequent irregular behaviour since 1968.

*) Revised version of a contribution to the 3th conference "Isotope in der Natur", Leipzig 1983

[1]) DDR-9200-Freiberg, Bernhard-von-Cotta-Straße 4

Studies of the content of tritium in Antarctic precipitation [7–9] have shown that since the commencement of nuclear weapons testing in the atmosphere there has also occured a marked increase of the tritium content in Antarctic precipitation, which is less pronounced than that observed in the northern hemisphere. Peak values were recorded around 1965, i.e., about two years [7, 8] later than in the northern hemisphere (Fig. 1).

Since the tritium contents of Antarctic precipitation were generally determined retrospectively from ice drill cores, in their evaluation consideration should be given not only to such influences as secular and regional variations in the rate of accumulation but to continental and altitudinal effects as well.

The object of this work is to describe, by a model representation, the secular and regional variations of tritium contents in precipitation in the northern and southern

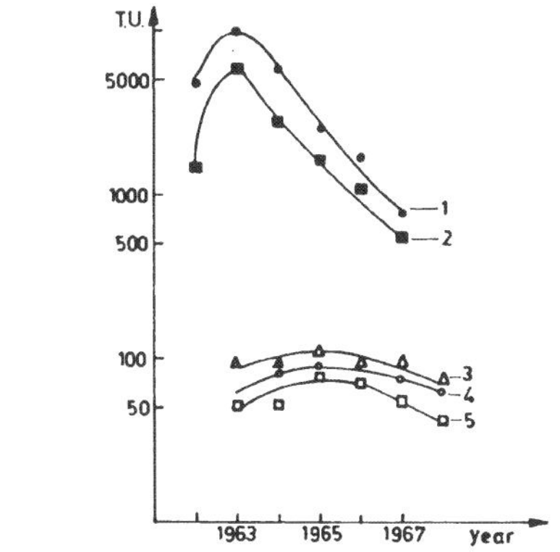

Fig. 1. Annual maximum values of tritium content of precipitation in northern and southern hemispheres [4]
1 – Nord; 2 – Ottawa; 3 – Adelaide/Melbourne; 4 – Falkland-Il.; 5 – Invercargill; New Zealand

hemispheres for the period after the cessation of nuclear weapons tests in the atmosphere.

2. Modelling the transport of tritium in the atmosphere

The model concept of *Nydal* [10] which has proved usefully in describing the global exchange of carbon dioxide incited the present author to study a multicompartment model of the essential stratospheric, tropospheric, and oceanic reservoirs that are involved in the global exchange of tritium (Fig. 2). As initial condition for the year 1963 it is assumed that tritium is released, in the form of a surge, in the stratosphere of the northern hemisphere and that the content of tritium in the other compartments of the model is comparatively low.

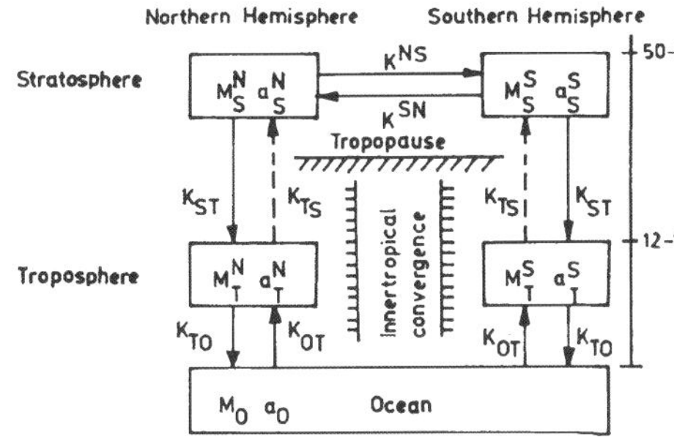

Fig. 2. Compartment model of global transport of tritium

For the stratosphere we have the following balances of tritium activites (A) and masses of moisture (M):

$$\dot A_S^N = -(K_{SS}^{NS} + K_{ST}^N + \lambda^{2)}) \cdot A_S^N + K_{SS}^{SN} \cdot A_S^S + K_{TS}^N \cdot A_T^N \quad (1)$$

$$\dot A_S^S = -(K_{SS}^{SN} + K_{ST}^S + \lambda) \cdot A_S^S + K_{SS}^{NS} \cdot A_S^N + K_{TS}^S \cdot A_T^S \quad (2)$$

$$\dot M_S^N = -(K_{SS}^{NS} + K_{ST}^N) \cdot M_S^N + K_{SS}^{SN} \cdot M_S^S + K_{TS}^N \cdot M_T^N \quad (3)$$

$$\dot M_S^S = -(K_{SS}^{SN} + K_{ST}^S) \cdot M_S^S + K_{SS}^{NS} \cdot M_S^N + K_{TS}^S \cdot M_T^S \quad (4)$$

If, wothout appreciable loss of generality it is assumed that

$$\left. \begin{array}{l} K_{SS}^{SN} = K_{SS}^{SN} = K_{SS}\,; \quad K_{ST}^N = K_{ST}^S = K_{ST}\,; \\ M_S^N = M_S^S \quad \text{and} \quad \dot M_S^N = \dot M_S^S = 0 \end{array} \right\} \quad (5)$$

then

$$K_{TS}^{N/S} = K_{ST}^{N/S} \cdot \frac{M_S}{M_T^{N/S}} \quad (6)$$

is obtained from eqs. (3) and (4).

Thus, the differential equations

$$\dot a_S^N = -(K_{SS} + K_{ST} + \lambda) a_S^N\, a + K_{SS} \cdot a_S^S + K_{ST} \cdot a_T^N \quad (7)$$

and

$$\dot a_S^S = -(K_{SS} + K_{ST} + \lambda)\, a_S^S + K_{SS} \cdot a_S^N + K_{ST} \cdot a_T^S \quad (8)$$

are obtained from eqs. (1) and (2) for the secular variation of the specific tritium activities $\left(a = \frac{A}{M}\right)$.

By taking into account that $a_T \ll a_S$, the following equations are obtained, from eqs .(7) and (8), for the secular variation of the specific tritium activity in the stratosphere

$$a_S^N(t) = C_1 \cdot e^{-(K_{ST}+\lambda)t} + C_2 \cdot e^{-(2K_{SS}+K_{ST}+\lambda)t} \quad (9)$$

and

$$a_S^S(t) = C_1 \cdot e^{-(K_{ST}+\lambda)t} - C_2 \cdot e^{-(2K_{SS}+K_{ST}+\lambda)t} \quad (10)$$

Here, t is the number of years after 1963.

The injection of tritium from the stratosphere into the troposphere occurs especially in the polar regions [1, 11, 12] and is accociated with the annual winter-to-summer-change of the distribution of pressure in the upper atmosphere [13, 14]. At Halley Bay Stations, for example, the maximum tritium levels recorded in the years 1966 to 1971 were observed in the monthly precipitations of August or September [3].

Therefore, it is assumed that the peak values of the tritium contents measured in polar precipitation are "characteristic" of the corresponding startospheric tritium level. If it is assumed that the injection of tritium from the stratosphere into the troposphere occurs once a year in the form of a surge and that tritium is subsequently distributed throughout the troposphere, then the annual peak values of the tritium contents in polar precipitation can be calculated, from the activity balance of the proportions of stratospheric and tropospheric moisture ($\alpha \cdot m_S$ and m_T, respectively, $\alpha < 1$) mixed during the annual injection, as

$$a_{pole}^{N/S}(t) = a_T^{N/S}(t-1) + \frac{\alpha \cdot m_S}{m_T} \cdot a_S^{N/S}(t) \quad (11)$$

As is evident from the decay behaviour of monthly tritium levels after the annual peak value (Fig. 1), tritium is very rapidly ($K_{TO} = 4{\cdot}15\ a^{-1}$ corresponding to $T_{1/2} - 2$ months) transferred from the troposphere to the ocean, so that, to a good approximation, the following equation may be regarded as holding true:

$$a_{pole}^{N/S}(t) = \frac{\alpha \cdot m_S}{m_T} \cdot a_S(t) \quad (12)$$

From eq. (12) it is apparent that the secular variation in the specific tritium activity of the stratosphere can be determined from polar precipitation. Accordingly, eqs. (9) and (10) also hold for the secular variation of annual peak values of tritium in tropospheric polar regions.

The parameters K_{ST}, K_{SS}, C_1, and C_2 are determined by fitting relations (9) and (10) to the measured results. Being typical of the northern and southern hemispheres, values measured at the Nord and Thule, Greenland, Stations [3] and the South Pole [7], respectively, were used. The chronology of the 1975 and 1978 snow pits of South Pole station,

[2]) λ-transformation constant of tritium $= 0{\cdot}0558\ a^{-1}$

given by *Jouzel* et al. [7], has been corrected as far as the year of highest tritium contents should be the year 1965, not 1966.

The coefficients of transport K_{SS} and K_{ST} are calculated from the equations

$$\left.\begin{aligned} a^N + a^S &= C_1 \cdot e^{-(K_{ST}+\lambda)t}\ ; \\ a^N - a^S &= C_2 \cdot e^{-(2K_{SS}+K_{ST}+\lambda)t} \end{aligned}\right\} \qquad (13)$$

see Fig. 3.

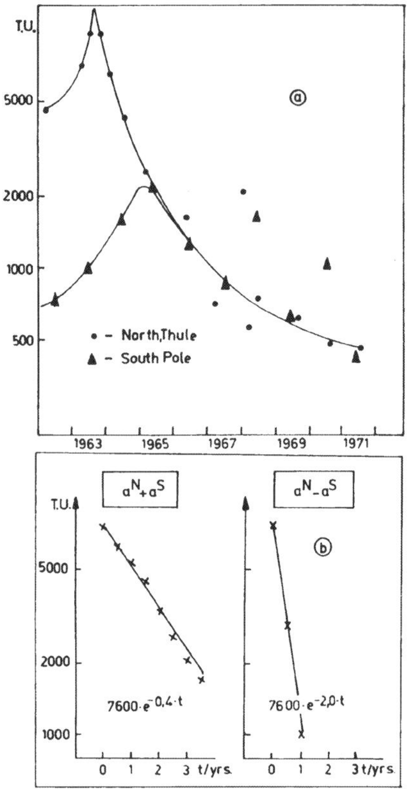

Fig. 3. a Annual maximum tritium contents in precipitation of northern and southern polar regions, taken from [4, 7]
b Evaluation of trnasport coefficients according eq. (13)

To the tropospheric tritium contents the following equations are obtained:

$$a_T^N(t) = 3800 \cdot e^{-0.40t} \cdot (1 + e^{-2.0t}) \quad \text{and} \qquad (14)$$

$$a_T^S(t) = 3800 \cdot e^{-0.40t} \cdot (1 - e^{-2.0t}) + 1200 \cdot e^{-4.15t} \qquad (15)$$

The results of model calculations together with the measured values are given in Fig. 4.

Summarized in Tab. 1 are the parameters determined from a comparison of eqs. (9) and (10), on the one hand, and eqs. (13) and (14), on the other.

By using the transport coefficients determined, we obtain, from the relation

$$t^S_{max} = \frac{1}{2K_{SS}} \ln\left(1 + \frac{2K_{SS}}{K_{ST} + \lambda}\right) \qquad (16)$$

$t^S_{max} = 1$ year. This is in agreement with the measured results reported by *Jouzel* et al. [7] and *Reinwarth* [8]. The tritium content of stratospheric moisture tends to decay with a half-life of $T_{1/2}$ (strat.) $= \ln 2/(K_{ST} + \lambda) = 20$ months cf. [25] and [7b].

A term had to be added to eq. (15) as used for the model calculations. This is the tritium level of the South Polar troposphere in 1963, which rapidly decays with a tropospheric residence time of 0·25 a.

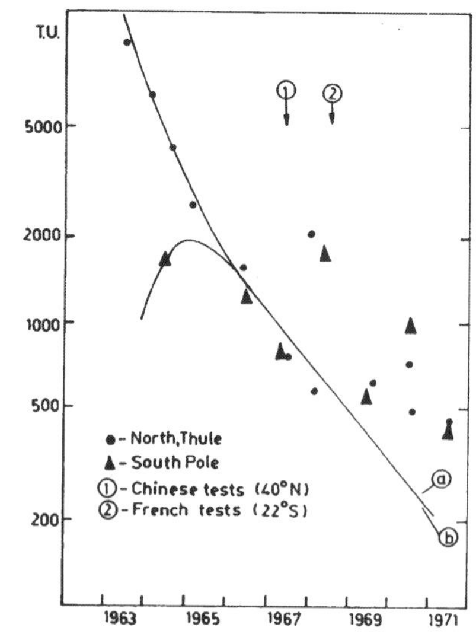

Fig. 4. Modelled variations of annual tritium peaks
a — North Polar region; b — South Polar region

Tab. 1. Atmospheric tritium transport parameters

Transition	Coefficient of transport
Beta transformation	$\lambda = 0{\cdot}056\ a^{-1}$
$S^N \to S^S$	$K_{SS} = 0{\cdot}80\ a^{-1}$
$S \to T$	$K_{ST} = 0{\cdot}35\ a^{-1}$
$T \to O$	$K_{TO} = 4{\cdot}15\ a^{-1}$

After 1969, thermonuclear tests again disturbed the modelled decay behaviour of tritium in precipitation. Nuclear weapons tests in 1968 performed by the People's Republic of China at 40 °C were found to be the cause of the peak recorded in 1969 in South Polar precipitation [7].

3. On the geographical variation of tritium contents in Antarctic precipitation

During the 21st Soviet Antarctic Expedition, samples of firn for isotope-physical analyses were collected in 1975/76 and 1977/78 from the Hays Glacier in the area of the Molodezhnaya (Enderby Land) Station [15]. Initial evaluations of tritium, deuterium, and ^{18}O analyses of these samples were presented in [16]. As shown by Fig. 5, a significant tritium peak is seen about 1965 with tritium level by 300 T.U. This is in agreement with results received from stations Halley Bay [3] and Filchner [8]. The mean tritium level recorded at Halley Bay Stations for the period from 1966 to 1971 is about 170 T.U. The mean tritium level recorded at Molodezhnaya Station for the period from 1965 to 1975 is of the order of 140 T.U. The level determined by *Jouzel* et al. for the South Pole (8/56—1/72) is about 550 T.U. [7].

The maximum tritium content of precipitation in the southern hemisphere of 2000 T.U. which can be calculated for 1965 from the present model was not observed in the coastal region of Antarctica, but could be found at the South Pole [7]. The causes of these differences can be found in the particular mechanism of injection and in the atmospheric circulation in Antarctica.

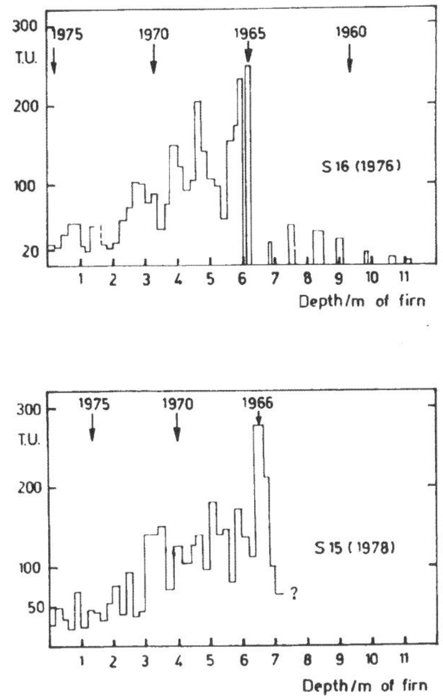

Fig. 5. Tritium in firn profiles of the Hays Glacier, Molodezhnaya (corrected to year of firn accumulation)

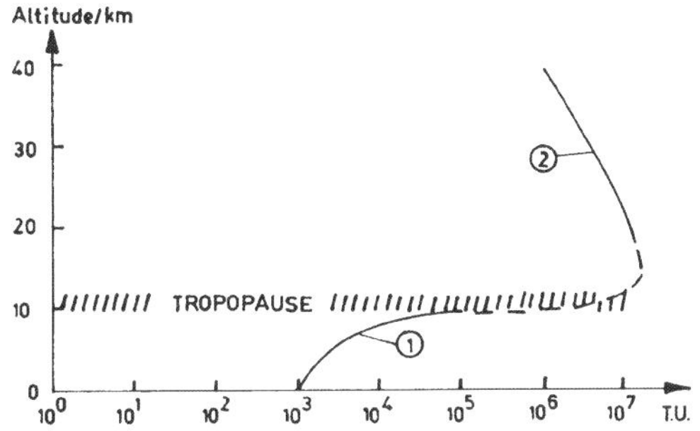

Fig. 6. Tritium in air moisture
1 – *Ehhalt*, 1966; 2 – *Pollock* et al., 1975

Ehhalt [17], *Mason* et al. [18], *O'Brien* et al. [19], and *Pollock* et al. [20] have made a number of studies to determine tropospheric and stratospheric tritium levels.

From the results obtained by those investigators it is apparent that the content of tritium in the stratosphere is about three orders of magnitude above that measured in the troposphere (Fig. 6).

From the results given by *Ehhalt* [17] it is possible to determine a doubling of the tritium content for the tropospheric region between 0 and 3000 metres of altitude. An altitudinal effect of about 100 per cent/3000 metres of increase in altitude should, therefore, be expected for tritium in Antarctica, too. Consequently, we have:

$$a(H) = a(H = 0) \cdot \left(1 + \frac{H}{3000\ \mathrm{m}}\right) \tag{17}$$

where H is the altitude above mean sea level and $a(H = 0)$ is the specific tritium activity, at sea level, on the (Antarctic) coast.

The results of measurements reported by *Jouzel* et al. for a Dumont D'Urville – Dome C – South Pole traverse (Fig. 7) are compatible with the results measured at the coastal stations referred to above, if the altitudinal effect is eliminated and the remaining difference from the mean tritium value represented as an "Antarctic effect" as a function of the distance from the South Pole (Tab. 2).

The difference $\bar{a} - a(H)$ was plotted in Fig. 8 as a function of the distance from the South Pole. It can be seen that a large area of the Antarctic inland up to a distance from the pole of 1700 kilometres is characterized by a (relatively uniformly) high mean tritium level ($\cong$ 400 T.U. for the 8/56 to 1/72 period), whereas the coastal region extending over a width of $\cong$ 500 kilometres showed only about 40 T.U. for the same period. The geographical regions clearly distinguished in Fig. 8 are, evidently, marked by the particular kind of air mass, and more specifically the air moisture.

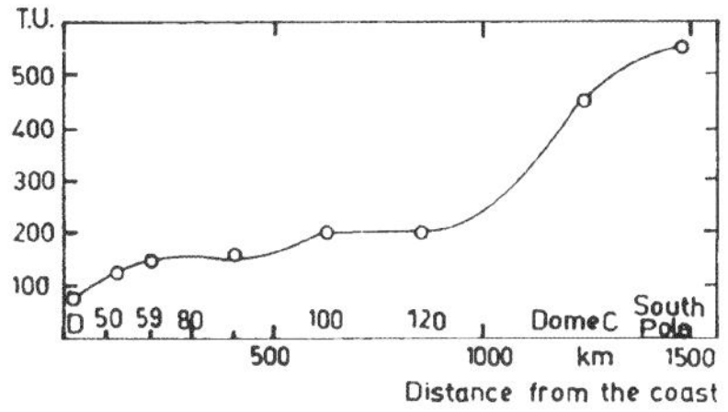

Fig. 7. Mean tritium levels for the period from 1956 to 1972 along a Dumont d'Urville – South Pole traverse (7a, 7b)

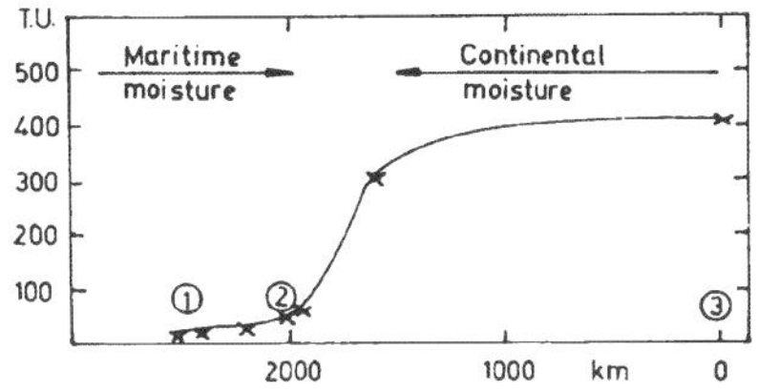

Fig. 8. Tritium continental effect in Antarctica
1 – Dumont d'Urville; 2 – Molodezhnaya; 3 – South Pole

Tab. 2. Antarctic Continental effect (values were taken from [7])

Position	$\bar{a}$/T.U.[1])	H/m above sea level	$a(H)$/T.U. (according to eq. (15) w. $a(H = 0)$ = 70 T.U.)	$\bar{a} - a(H)$/T.U.	Distance from the South Pole/ kilometres
D 41	80	300	90	−10	2550
D 50	125	1500	105	20	2500
D 59	150	2100	119	31	2400
D 80	160	2500	128	32	2200
D 100	200	2800	135	65	200
D 120	200	3000	140	60	1950
Dome C	450	3300	147	303	1600
South Pole	550	2800	135	415	0

[1]) Mean value for the 1956 to 1972 period

Whereas in the interior of the Antarctic continent, because of the meteorological conditions prevailing there, air masses originating from higher tropospheric strata and enriched with tritium through stratospheric injection are predominant, maritime and low-tritium air masses are prevalent in a coastal strip of several hundred kilometres because of the circum-Antarctic cyclone belt [21].

Because of admixture of moist maritime air the Antarctic continental effect of tritium results in the tritium content being reduced with increasing distance from the pole. The decrease to ine-tenth of the value of the inner Antarctic region starts abruptly at $\cong$ 1600 kilometres. Since the rate of accumulation in the coastal region of Antarctica is increased by a factor of 4 [7, 16, 23] as compared with the polar region, the tritium activity balance shows that, compared to the coastal area, approximately 2·5 times as much stratospheric air moisture is injected in the polar area. This supports the concept of tritium injection into the troposhere occurring in the near-pole region (i.e., beyond 70 °S) [1].

The Antarctic continental effect can also be used to explain the deviation of ^{18}O levels (measured on precipitation at coastal stations) from values calculated from the corresponding average annual temperature as being due to the admixture of moist maritime air [24].

Thus, the Antarctic continental effect of tritium described here is basically different from the continental effect observed in the northern hemisphere [22], which involves an increase in tritium content with increasing distance from the Atlantic coast because of the admixture to maritime air masses of masses of moist reevaporating higher-tritium air.

Acknowledgements

The author is indebted to the members of the GDR groups in the 21st and 23rd S.A.Es, especially to *D. Tauchert*. His thanks are also due to Prof. Dr. *K. Fröhlich*, head of the "Natural Radionuclides/Environmental Physics" working group, as well as to workers in the Tri-Car-Laboratory, Department of Physics, Freiberg Mining Academy.

Received September 6, 1984
Accepted in revised form November 15, 1984

References

[1] *W. R. Schell, G. Sauzay, B. R. Payne*, Proc. IAEA-SM-181/34, IAEA, Vienna 1974, p. 375
[2] *W. Roether*, Isotopes in Hydrology. IAEA, Vienna 1967, p. 73
[3] Environmental Isotope Data, Nos. 1–6, IAEA, Vienna 1969, 1970, 1971, 1973, 1975, 1979
[4] *R. M. Brown*, Isotope Hydrology. Proc. IAEA-SM-129/1, IAEA, Vienna 1970, p. 3
[5] *N. Esser*, Thesis, Heidelberg 1980
[6] *K. Fröhlich, D. Hebert, H. Jordan*, Isotopenpraxis **14** (1978) 387
[7a] *J. Jouzel, M. Pourchet, C. Lorius, L. Merlivat*, Proc. IAEA-SM-232/38, IAEA, Vienna 1978, p. 31
[7b] *J. Jouzel, L. Merlivat, M. Pourchet, C. Lorius*, Earth and Planetary Science Letters **45** (1979) 61
[8] *O. Reinwarth*, Polarforschung **51** (1) (1981) 61
[9] *D. Hebert, K. Fröhlich*, ZfI-Mit. **30** (1980) 305
[10] *R. Nydal*, Journal of Geophys. Res. **73** (1968) 3617
[11] *E. R. Reiter*, Proc. IAEA-SM-181/40, IAEA, Vienna 1974
[12] *H. Flohn*, Quarternary Research **12** (1979) 135
[13] *K. Schneider-Carius*, Die Grundschicht der Troposphäre. Leipzig: Akad. Verl.-Ges. Geest & Portig KG 1953, p. 94
[14] *G. Skeib*, Antarktika. Leipzig – Jena – Berlin: Urania-Verl. 1965, p. 40
[15] *R. Eger*, et al., Geod.-Glaziol. Arbeit am Hays-Gletscher. Geod.-Geophys. Veröff. R III, **49** (1981), Berlin 1981
[16] *D. Hebert, K. Fröhlich*, ZfI-Mitt. 51 (1982) 87
[17] *D. H. Ehhalt*, Journal of Geophys. Res. **76** (1971) 30, 7351
[18a] *A. S. Mason, G. Östlund*, Journal of Geophys. Res. **81** (1976) 30, 5349
[18b] *A. S. Mason, G. Östlund*, Proc. IAEA-SM-232/62, IAEA, Vienna 1978
[19] *R. S. O'Brien, W. F. J. Evans*, Journal of Geophys. Res. **86** (1981) C 12, 12101
[20] *W. Pollock, L. E. Heidt, R. Lueb, D. H. Ehhalt*, Journal of Geophys. Res. **85** (1980) C 10, 5555
[21] *A. Helbig*, Geod.-Geophys. Veröff. R I (1980) 7, 64
[22] *W. Weiss, J. Bullacher, W. Roether*, Proc. IAEA-SM-232/18, IAEA, Vienna 1978
[23] *S. Orvig*, Climates of the Polar Regions. Amsterdam – London – New York: Elsevier Publ. Company 1970
[24] *E. Eriksson*, Guidebook on Nuclear Techniques in Hydrology (ed. 83), IAEA, Vienna 1983
[25] *A. S. Mason, G. Hut, K. Telegadas*, Tellus **34** (1982) 369

Isotopenpraxis **21** (1985) 8, S. 272

Buchbesprechung

Radiation Protection and Measurement for Low Voltage Neutron Generators

NCRP Report No. 72 (Issued November 1, 1983), 80 S., div. Abb. u. Tab.; kart.

Der Report des National Council on Radiation Protection and Measurements der USA behandelt Fragen des Strahlenschutzes und der Meßtechnik an Niedervolt-Neutronengeneratoren. Unter diesen Generatoren werden solche verstanden, die unterhalb einiger 100 kV arbeiten und Neutronen primär durch die T(d, n)-Reaktion erzeugen. Da die Empfehlungen für derartige Generatoren über eine Reihe von NCRP-Reports verstreut sind, hat man sich zu dieser Zusammenfassung aller relevanten Empfehlungen entschlossen. Der Report ist also gedacht als Anleitung zu einer guten praktischen Handhabung von Niedervolt-Neutronengeneratoren. Dieses Anliegen wird auch aus der Gliederung verständlich:

Nach Behandlung der durch Niedervolt-Neutronengeneratoren erzeugten Strahlungsarten (Neutronen, Röntgen- und Gammastrahlen, induzierte Radioaktivität) werden die Grundlagen des Strahlenschutzes dargelegt (Kontrolle der Strahlenexposition, Strahlenschutzgrenzwerte, Personenkontrolle, Überschreitung von Grenzwerten). Es schließen sich Ausführungen zu Abschirmungen und anderen physischen Strahlenschutzmaßnahmen und der Strahlenschutzmeßtechnik an (insbesondere zur Messung von Tritium). Schließlich werden Quellen, Natur, Behandlung und Lagerung von radioaktiven Abfällen beschrieben.

Abschließend werden andere Gefährdungsmöglichkeiten angegeben (Hochspannung, Implosion von evakuierten Glaskomponenten, gasförmige Noxen) sowie Aussagen zum Genehmigungsverfahren (in den USA) gemacht.

Der Anhang enthält Begriffsdefinitionen, Methoden der Tritiummessung, DT-Neutronenabschirmungen (Transmissionskurven für 14 MeV-Neutronen in Wasser und Beton) sowie ein Statement des NCRP zum Dosislimit für Neutronen (vom 21. 2. 1980).

K. Hübner

Isotopenpraxis 21 (1985) 8, S. 273 bis 277

Zur Strahlenchemie des heterogenen Systems Trichlortrifluorethan—Arsen

U. Prösch, L. Tepper

(Humboldt-Universität zu Berlin, Sektion Chemie)[1]

Entgaste Suspensionen von Arsen in Trifluortrichlorethan wurden mit einer ^{60}Co-Bestrahlungsanlage bestrahlt. Anschließend wurden die Proben mit Natriumhydroxidlösung behandelt und die Chlorid- und Fluoridausbeuten bestimmt. Die Ausbeute von $C_4F_6Cl_4$ wurde quantitativ gaschromatographisch gemessen, andere Perhalogenkohlenwasserstoffe qualitativ beobachtet. Die Konzentrationen an Fluorid und Chlorid stiegen linear dosisproportional. Die Abhängigkeit der Ausbeuten an Fluorid, Chlorid und $C_4F_6Cl_4$ in Abhängigkeit von der Menge Arsen bei konstanter Zugabe von $C_2F_3Cl_3$ wurde untersucht.

Deaerated suspensions of arsenic in trichlorotrifluoroethane were irradiated in a ^{60}Co radiation facility. After treating the irradiated samples with sodium hydroxide solution the yields of chloride and fluoride were determined. By gaschromatography the yield of $C_4F_6Cl_4$ was measured quantitatively. The yields of chloride and fluoride proved to be linear functions of the applied radiation dose. The dependence of the chloride, fluoride and $C_4F_6Cl_4$ yields from the amount of arsenic added to a fixed quantity of $C_2F_3Cl_3$ was studied. The experimental results were interpreted by a radical reaction mechanism.

Keywords

arsenic; cobalt 60; organic halogen compounds; radiation chemistry; reaction kinetics

1. Einleitung

Die nichtthermische Energieeinwirkung (UV-, ionisierende Strahlung, elektrische Gasentladung) liefert in Perhalogenkohlenwasserstoffen Halogenatome und halogenorganische Radikale. Ihre Reaktion mit Festkörperoberflächen ist von Interesse bezüglich der physikalischen und chemischen Modifizierung von Oberflächenschichten oder -bereichen, einschließlich des vollständigen Aufbaus oder Abtrags solcher Schichten. So befindet sich gegenwärtig z.B. das plasmachemische Ätzen bei der Strukturierung von mikroelektronischen Bauelementen in rascher Entwicklung und technischer Realisierung.

Die chemischen Vorgänge bei der Einwirkung nichtthermischer Energie auf heterogene Systeme sind durch ein kompliziertes Zusammenwirken von Parallel- und Folgereaktionen unter den Bedingungen einer reaktiven Beteiligung der Festkörperoberfläche gekennzeichnet. Eine qualitativ und quantitativ ausreichende Beschreibung solcher Reaktionsmechanismen liegt gegenwärtig nur in begrenztem Maße vor und kann voraussichtlich erst schrittweise durch die Aufklärung von Teilaspekten und vergleichende Betrachtung der Ergebnisse an unterschiedlichen stofflichen Systemen und bei Verwendung unterschiedlicher Energieformen erarbeitet werden. Einige Ergebnisse mit orientierendem Charakter hinsichtlich der Methodik und Interpretationsmöglichkeiten am System Trichlortrifluorethan—Arsen sollen als Beitrag zu diesem Themenkreis mitgeteilt werden.

Für die Systemauswahl lagen mehrere Gründe vor. An erster Stelle stand der Gesichtspunkt, daß im Trichlortrifluorethan $CF_2Cl-CFCl_2$ (nachfolgend abkürzend und unvollständig als R113 bezeichnet) eine gut handhabbare, in größerem Umfang produzierte Verbindung vorliegt, die unter Einwirkung nichtthermischer Energie sowohl Chlor- als auch Fluorradikale liefert, deren unterschiedliche Wirkung in plasmachemischen Prozessen diskutiert wird. Außerdem ist noch die dritte wichtige Grundreaktion in der Klasse der Perhalogenkohlenwasserstoffe, die Kohlenstoff-Kohlenstoffbindungsspaltung, zu beobachten, so daß alle hauptsächlich zu erwartenden Reaktionsarten dieser Stoffklasse verfolgt werden können.

In bezug auf den Festkörper wurde Arsen gewählt, da für mikroelektronische Bauelementzwecke dieses Element sowohl als Dotierungskomponente wie auch als Hauptsubstratbestandteil z.B. in GaAs eine wesentliche Rolle spiel

Die Verwendung von ^{60}Co-Gammastrahlung als nichtthermische Energieform zum Studium des Systems ergab sich aus dem Bestreben, primär möglichst jeweils nur eine Bindung je Molekül zu spalten und keinen nichtthermischen Folgeumsatz von Radikalkombinationsverbindungen oder arsenhaltigen Produkten zu bewirken. Diese Forderung kann bei strahlenchemischer Reaktionsführung gut erfüllt werden. Untersuchungen von *Tominaga* und Mitarbeitern [1, 2] und von *Prösch* und Mitarbeitern [3—5] über die Strahlenchemie von reinem R113 und in Verbindung mit Scavengern ergaben, daß neben der C—Cl-, C—C- und C—F-Bindungsspaltung allenfalls eine simultane Freisetzung von Cl_2 eintritt, gemäß dem Primärreaktionsschema:

$$C_2F_3Cl_3 \rightsquigarrow C_2F_3Cl_2^{\cdot} + Cl^{\cdot} \quad (1)$$

$$C_2F_3Cl_3 \rightsquigarrow CF_2Cl^{\cdot} + CFCl_2^{\cdot} \quad (2)$$

$$C_2F_3Cl_3 \rightsquigarrow C_2F_2Cl_3^{\cdot} + F^{\cdot} \quad (3)$$

$$C_2F_3Cl_3 \rightsquigarrow C_2F_3Cl + Cl_2 \quad (4)$$

Bei Bestrahlungsdosen von 10^4 bis 10^6 Gy werden gemäß den unterschiedlichen *G*-Werten Konzentrationen der Radikalkombinationsprodukte von 10^{-3} bis 10^{-1} M erzielt, deren strahlenchemischer Folgeumsatz vernachlässigbar klein ist. Demgegenüber ist in Perhalogenkohlenwasser-

[1]) Anschrift: 1040 Berlin, Hessische Str. 1—2

stoff—GaAs-Systemen bei plasmachemischer Reaktionsführung von uns beobachtet worden [6], daß eine vollständige Chlorfreisetzung und eine 30—40%ige Fluorfreisetzung aus den Ausgangsverbindungen eintritt, woraus auf einen intensiven Radikal- und Molekülabbau in sekundären Reaktionen zu schließen ist, der die Verfolgung und Beschreibung der Reaktionsmechanismen zusätzlich kompliziert.

2. Experimentelles

Die Verbindung $CF_2Cl-CFCl_2$ wurde als Fridohna 113 (VEB Fluorwerke Dohna) bezogen. Arsengranulat (VEB Laborchemie Apolda) wurde gemörsert und gesiebt. Für die Versuche wurde eine Siebfraktion < 50 µm verwendet. Nach mikroskopischen Aufnahmen und einer Sedimentationsanalyse lag der mittlere Durchmesser der Teilchen bei $d_{As} = 31$ µm. Es wurden jeweils $v = 1{,}5$ ml R113 mit 0 bis 2,5 g Arsenpulver in kleinen Kölbchen entgast (3 Frier-Tau-Zyklen) und abgeschmolzen. Die Proben wurden in einer Gamma MRCH-100-^{60}Co-Bestrahlungsanlage bei einer Dosisleistung von $I = 117$ Gy/min bestrahlt. Die Bestrahlungszeiten variierten zwischen 1 und 12 Tagen. Nach der Bestrahlung erfolgte eine Analyse der nicht mehr organisch gebundenen Halogenanteile oder eine gaschromatographische Untersuchung. Zur Halogenanalyse wurden die Kölbchen mit den bestrahlten Proben in einer verschlossenen Pulverflasche, die 40 ml 0,1 M NaOH enthielt, durch kräftiges Schütteln zerbrochen. Nach einigen Minuten wurden das Arsenpulver und die Glasanteile abfiltriert und ausgewaschen. Die alkalische Lösung und die Waschwasseranteile wurden vereinigt und in einem Maßkolben auf 100 ml aufgefüllt. Zur Analyse wurden Aliquots mit 2 M HNO_3 neutralisiert. Die Chloridbestimmung erfolgte potentiometrisch mit 0,01 M $AgNO_3$, während die Fluoridbestimmung mit einer ionenselektiven Elektrode vorgenommen wurde, die mit eingestellten NaF-Standardlösungen im Bereich von 10^{-1} bis 10^{-4} M Fluorid geeicht wurde.

Die gaschromatographischen Analysen wurden mit einem Gerät GCHF 18.3 des VEB Chromatron ausgeführt. Als Trennsäulenmaterial diente 10% Methylsilikonöl NM 1-1000 auf Chromosorb P. Die Trennsäulenlänge betrug 3 m. Als Schleppgas wurde Wasserstoff bei einem Volumenstrom $v = 60$ ml/min benutzt. Der Substanznachweis erfolgte mit einem Wärmeleitfähigkeitsdetektor. Die Substanztrennung wurde nichtisotherm vorgenommen. Nach einem isothermen Vorlauf von 4 min bei 30 °C wurde die Temperatur bei einer Heizrate $r = 6$ °C/min auf 140 °C gesteigert.

In Abb. 1 ist das Gaschromatogramm einer bestrahlten Probe dargestellt. Die halogenorganischen Reaktionsprodukte wurden in einer GC-MS-Anlage an einem gleichartigen Säulenmaterial identifiziert[2]). Zur quantitativen Analyse von $C_4F_6Cl_4$ (R316) wurde jeweils 1 ml bestrahlter Probe mit 20 µl Heptan versetzt und 5 µl des Gemisches injiziert. Die molare Konzentration des R316 in der bestrahlten Probe ergab sich empirisch aus den gemessenen Bergflächen zu

$$c_{R316} = 0{,}135 \cdot F_{R316}/F_{Heptan}$$

3. Ergebnisse und Diskussion

Bei der Bestrahlung der Aufschlämmung von Arsen in R113 werden Halogenatome gemäß den Gleichungen (1), (3) und evtl. (4) freigesetzt. Durch Radikalkombinationen bilden sich Cl_2, F_2 und ClF. Außerdem werden Halogenatome an der Arsenoberfläche gebunden, was durch die Bezeichnung $\{As_xCl\}$ bzw. $\{As_xF\}$ symbolisiert sei. Bei genügend langer Reaktionszeit ist mit der Bildung von $AsCl_3$ und AsF_3 zu rechnen, die als anorganische, covalente Halogenide bei Zimmertemperatur flüssig und mit R113 begrenzt mischbar sind. Bei der anschließenden Aufarbeitung der bestrahlten Proben mit Natronlauge erfolgt Hydrolyse. Die Halogenmoleküle und Interhalogen, sofern sie nicht in thermischer Nachreaktion ebenfalls $AsCl_3$ und AsF_3 gebildet haben, sollten bei der Hydrolyse neben Halogenidionen auch Hypochlorit liefern. Wegen der Gegenwart von Arsen und Arsenit wird dieses jedoch sofort reduziert. Somit ist als strahlenchemische Ausbeutebilanz zu erwarten:

$$G(Cl^-) = G(\{As_xCl\}) + 2G(Cl_2) + G(ClF)$$
$$G(F^-) = G(\{As_xF\}) + 2G(F_2) + G(ClF)$$

Bei Proben konstanter Zusammensetzung, z.B. 2 ml R113 und 2 g Arsenpulver gleicher Siebfraktion sollten die Chlorid- und die Fluoridkonzentrationen linear dosisproportional ansteigen, wenn Radikalkombinationen und Radikalabfang durch Arsen nicht durch Folgereaktionen beeinträchtigt werden. In Abb. 2 und 3 sind die in Abhängigkeit von der Dosis gemessenen Konzentrationen dargestellt. Im Rahmen der Fehlergrenzen, die bei den Fluoridbestimmungen mit durchschnittlich 10% und bei den Chloridanalysen mit 5% anzusetzen sind, existiert diese Linearität. Die lineare Regressionsrechnung zeigte, daß die Geraden nicht durch den Koordinatenanfangs-

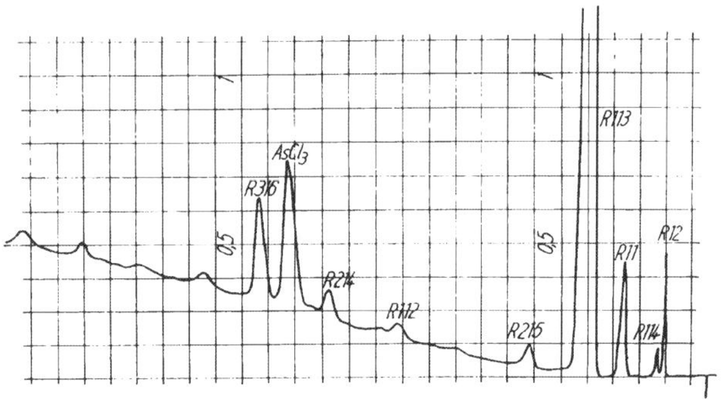

Abb. 1. Gaschromatogramm von R113/As-Aufschlämmungen (1,5 ml/2 g) nach Bestrahlung mit $2 \cdot 10^5$ Gy

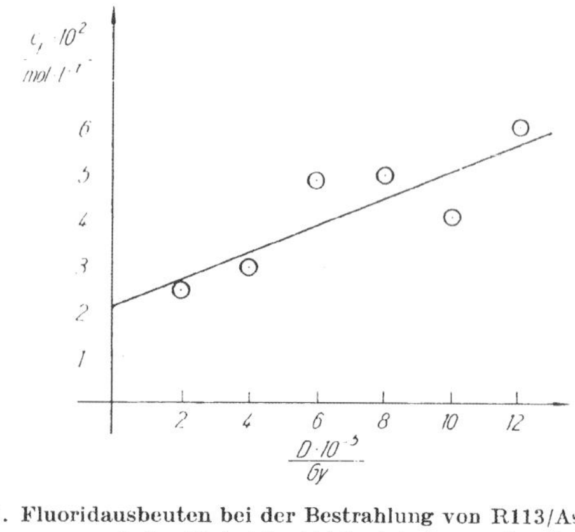

Abb. 2. Fluoridausbeuten bei der Bestrahlung von R113/As-Aufschlämmungen (1,0 ml/2,4 g) in Abhängigkeit von der Strahlendosis

[2]) Für die Durchführung und Interpretation der GC-MS Analysen sei Dr. *L. Alder* (Sektion Chemie der HUB) gedankt

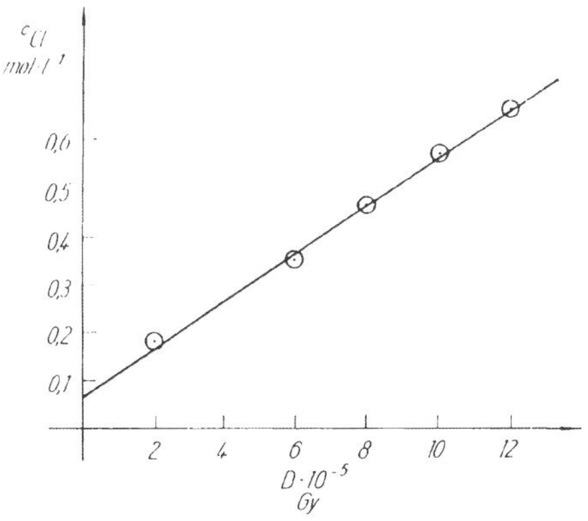

Abb. 3. Chloridausbeuten bei der Bestrahlung von R113/As-Aufschlämmungen (1,0 ml/2,4 g) in Abhängigkeit von der Strahlendosis

punkt gehen, d. h., daß ein Blindwert von Halogenidionen zu berücksichtigen ist, der seine Ursachen in einer geringen thermischen oder photochemischen Zersetzung des R113 bei Lagerung oder einer geringen Hydrolyse des R113 durch die alkalische Aufarbeitung der Proben haben kann. Die Resultate bestätigen, daß im gewählten Dosisbereich keine komplizierenden strahleninduzierten Folgereaktionen auftreten. Demgemäß kann das Reaktionsgeschehen durch folgendes Modell dargestellt werden:

Als Primärreaktionen werden die Reaktionen (1) bis (3) berücksichtigt. Damit liegen 6 Radikalarten vor. Ihre Ausbeuten sind durch die Scavengeruntersuchungen bei der Radiolyse homogener Systeme (l. c.) in erster Näherung bekannt. Für die weiteren Berechnungen wurde angenommen:

$$G_{C_2F_3Cl_2^{\cdot}} = G_{Cl^{\cdot}} = 4{,}9$$

$$G_{CF_2Cl^{\cdot}} = G_{CFCl_2^{\cdot}} = 1{,}4$$

$$G_{F^{\cdot}} = G_{C_2F_2Cl_3^{\cdot}} = 0{,}9$$

Prinzipiell sind nunmehr alle Radikale zur Kombination miteinander befähigt. Zur Vereinfachung wurde angenommen, daß $CF_2Cl^{\cdot}$ und $CFCl_2^{\cdot}$. bezüglich ihres Reaktionsvermögens genügend ähnlich sind, um sie gegenüber $C_2F_3Cl_2^{\cdot}$, $C_2F_2Cl_3^{\cdot}$ oder den Halogenatomen einheitlich zu behandeln. Es wurde deshalb mit fünf Zwischenprodukten gerechnet, die in folgender Weise bezeichnet wurden:

$$R_1 \triangleq C_2F_3Cl_2^{\cdot}\ ; \qquad R_2 \triangleq Cl^{\cdot}\ ;$$

$$R_2 \triangleq CF_2Cl^{\cdot} + CFCl_2^{\cdot}\ ; \qquad R_4 \triangleq C_2F_2Cl_3^{\cdot}$$

$$R \triangleq F^{\cdot}$$

Die Reaktionen des Radikalumsatzes sind damit durch 15 Geschwindigkeitsgleichungen erfaßbar:

$$l = 1, 2, \ldots 15$$

$$i = 1, 2, \ldots, 5\ ; \qquad j = i, i+1, \ldots 5\ ;$$

$$y_l \triangleq \left(\frac{\mathrm{d}c_{R_i}}{\mathrm{d}t}\right)_l = k_l \cdot c_i \cdot c_j$$

Dem Radikalumsatz durch Kombination sind die Abfangreaktionen mit Arsen hinzuzufügen:

$$Cl^{\cdot} + \{As_x\} \rightarrow \{As_xCl\} \tag{5}$$

$$F^{\cdot} + \{As_x\} \rightarrow \{As_xF\} \tag{6}$$

Die Reaktionsgeschwindigkeit dieser beiden Prozesse wurde nach dem ersten *Fick*schen Gesetz berechnet unter der Annahme, daß ein Halogenatom X bei jedem Stoß auf die Arsenoberfläche chemisch fixiert wird. Somit ist die Konzentration an freien Radikalen bei dem Abstand $x = 0$ von der Oberfläche $c_0 = 0$. Bei einer Diffusionsschicht der Dicke d_S innerhalb derer die Konzentration linear auf den Wert in der bestrahlten Flüssigkeit c_X ansteigt, ergibt sich für ein Flüssigkeitsvolumen v und eine Arsenoberfläche F:

$$\frac{\mathrm{d}n_X}{\mathrm{d}t} = -\frac{D_X \cdot F}{d_S} \cdot \left(\frac{\mathrm{d}c_X}{\mathrm{d}x}\right)_0$$

$$\frac{\mathrm{d}c_X}{\mathrm{d}t} = -\frac{D_X \cdot F}{d_S \cdot v} \cdot (c_X - c_0) = -\frac{D_X \cdot F}{d_S \cdot v} \cdot c_X$$

Den Reaktionen des Radikalverbrauches stehen während der Bestrahlung die strahleninduzierten Radikalbildungen gegenüber. Wenn I in Gy/min gemessen wird und mit $\varrho = 1{,}56$ die Dichte des R113 bezeichnet wird, so ist:

$$B_R^{\cdot} \triangleq \frac{\mathrm{d}c_R}{\mathrm{d}t} = G_R^{\cdot} \cdot I \cdot \varrho \cdot 1{,}73 \cdot 10^{-9}\ \mathrm{mol} \cdot \mathrm{l}^{-1} \cdot \mathrm{s}^{-1}$$

Für die fünf Radikalarten ergeben sich somit die folgenden Differentialgleichungen:

$$\frac{\mathrm{d}c_{C_2F_3Cl_2}}{\mathrm{d}t} = B_{R_1} - \sum_{i=1}^{5} y_i$$

$$\frac{\mathrm{d}c_{Cl}}{\mathrm{d}t} = B_{R_2} - (y_2 + y_6 + y_7 + y_8 + y_9) - \frac{D_{Cl} \cdot F}{d_S \cdot v} \cdot c_{Cl}$$

$$\frac{\mathrm{d}c_{R_3}}{\mathrm{d}t} = 2 \cdot B_{R_3} - (y_3 - y_7 - y_{10} - y_{11} + y_{12})$$

$$\frac{\mathrm{d}c_{R_4}}{\mathrm{d}t} = B_{R_4} - (y_4 + y_8 + y_{11} + y_{13} + y_{14})$$

$$\frac{\mathrm{d}c_{R_5}}{\mathrm{d}t} = B_{R_5} - (y_5 + y_9 + y_{12} + y_{14} + y_{15}) - \frac{D_F \cdot F}{d_S \cdot v} \cdot c_F$$

Das System der Differentialgleichungen kann nach einem numerischen Verfahren (z.B. *Runge-Kutta*) gelöst werden. Dabei wird das Zeitintervall so weit ausgedehnt, bis alle Radikalkonzentrationen stationäre Werte angenommen haben. Dieser Zustand ist nach dem *Bodenstein*-Prinzip zu erwarten. Daher kann auch unter der Voraussetzung $\mathrm{d}c_R/\mathrm{d}t = 0$ das entsprechende System algebraischer Gleichungen bezüglich der stationären Radikalkonzentrationen direkt gelöst werden. Wegen der Eigenschaften des vorliegenden Gleichungssystems wurde bei dieser Vorgehensweise ein auf der Methode der *regula falsi* basierendes, aber erweitertes Rechenverfahren von *Paulin* und *Griepentrog* [7] angewendet.

Zur Durchführung der Berechnungen waren Werte für die Diffusionskoeffizienten der Halogenatome, die Oberfläche des Arsens, die Diffusionsschichtdicke und die Geschwindigkeitskonstanten vorzugeben. Die Oberfläche des Arsens wurde für kugelförmige Teilchen errechnet:

$$F = \frac{6}{d_{As} \cdot \varrho_{As}} \cdot m_{As}$$

Die Dichte des Arsens beträgt 5,72. Die Arsenmenge m_{As} wurde zwischen 0,2 und 2,5 g je Probe variiert. Die Diffusionsschichtdicke $d_S = 5 \cdot 10^{-5}$ cm, als freier Parameter geschätzt, erwies sich als günstig. Die Reaktionsgeschwindigkeitskonstanten der einzelnen Reaktionen wurden summarisch betrachtet und beim Testen des Modells gruppenweise in größeren Intervallen variiert.

Das reaktionskinetische Modell wurde mit den experimentellen Werten für die Konzentration von Chlorid, Fluorid und $C_4F_6Cl_4$ in Abhängigkeit von der zugesetzten Arsenmenge bei einer konstanten Dosis von $4 \cdot 10^5$ Gy und 1,5 ml R133 je Probe verglichen. In Abb. 4 bis 6 sind die mittleren Konzentrationen aus je 6 Einzelproben bei einer gegebenen Arsenmenge dargestellt und die aus dem Modell berechneten Kurvenverläufe eingezeichnet.

Bezüglich der Halogenkonzentrationen ist zu beachten, daß sich diese aus den Anteilen des direkten Abfanges und der Erfassung molekularen Halogens zusammensetzen.

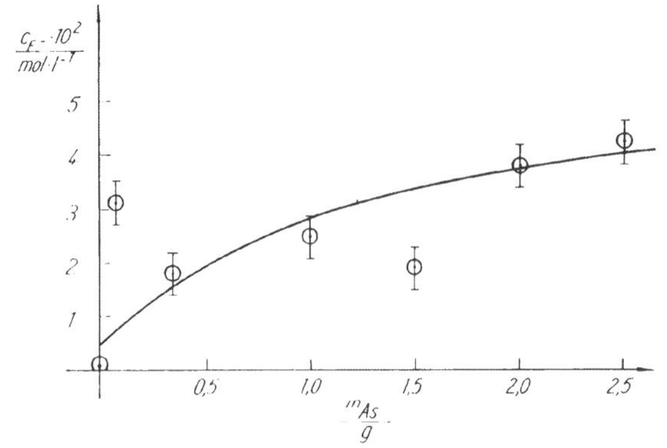

Abb. 4. Fluoridausbeuten bei der Bestrahlung von R113/As-Aufschlämmungen bei konstanter Dosis von $4 \cdot 10^5$ Gy in Abhängigkeit von der Arsenmenge je 1,5 ml R113

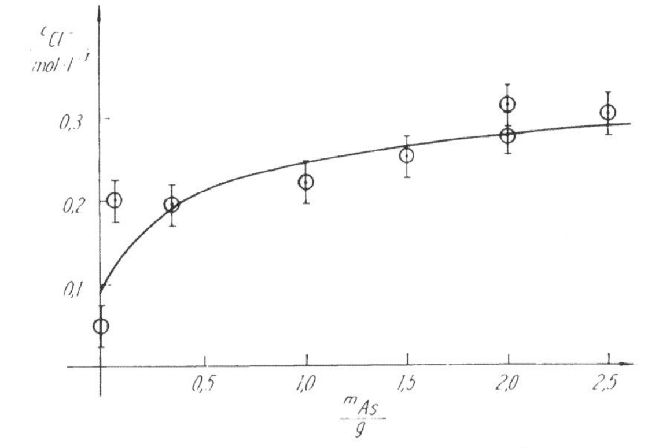

Abb. 5. Chloridausbeuten bei der Bestrahlung von R113/As-Aufschlämmungen bei konstanter Dosis von $4 \cdot 10_5$ Gy in Abhängigkeit von der Arsenmenge je 1,5 ml R113

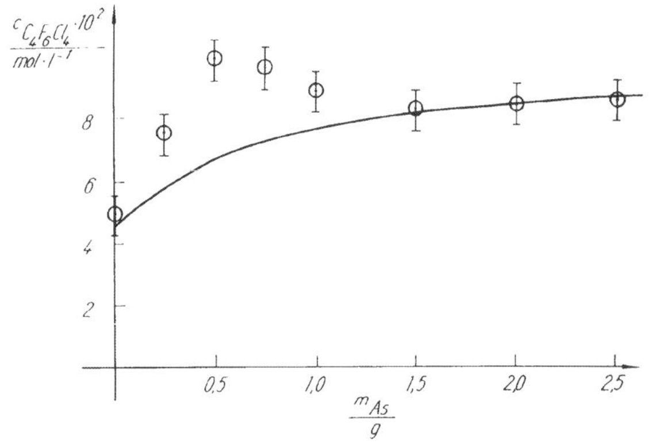

Abb. 6. Ausbeuten von $C_4F_6Cl_4$ bei der Bestrahlung von R113/As-Aufschlämmungen bei konstanter Dosis von $4 \cdot 10_5$ Gy in Abhängigkeit von der Arsenmenge je 1,5 ml R113

Tab. 1. Werte der Reaktionsgeschwindigkeitskonstanten der Radikalkombinationen im kinetischen Modell in l/mol · s

$k(R_1-R_1)$	$k(R_1-Cl)$	$k(R_1-R_3)$	$k(R_1-R_4)$	$k(R_1-F)$
10^9	$1{,}5 \cdot 10^9$	10^9	10^9	$2 \cdot 10^9$
$k(Cl-Cl)$	$k(Cl-R_3)$	$k(Cl-R_4)$	$k(Cl-F)$	$k(R_3-R_3)$
10^9	10^9	10^9	$2 \cdot 10^9$	$2 \cdot 10^9$
$k(R_3-R_4)$	$k(R_3-F)$	$k(R_4-R_4)$	$k(R_4-F)$	$k(F-F)$
10^9	$2 \cdot 10^9$	10^9	$2 \cdot 10^9$	$5 \cdot 10^9$

Der theoretische Wert für die Konzentration von R316 ergibt sich bei einer Bestrahlungszeit t zu:

$$c_{R316} = 0{,}5 \cdot k_1 \cdot (c_{R_1})_s^2 \cdot t$$

wobei $(c_{R_1})_s$ die stationäre Konzentration von R_1 ist.

In Tab. 1 sind die verwendeten Reaktionsgeschwindigkeitskonstanten angegeben. Sie liegen in der Größenordnung 10^9 l/mol · s und damit im Erwartungsbereich für einfache Radikalreaktionen. Die Kombinationen des kleinen Fluoratoms mit anderen Radikalen und insbesondere die Fluormolekülbildung wurden als die schnellsten Reaktionen angesehen. Es wird außerdem davon ausgegangen, daß die Rekombinationsreaktionen zum Ausgangsmolekül:

$$C_2F_3Cl_2^{\cdot} + Cl^{\cdot} \rightarrow C_2F_3Cl_3 \quad (7)$$

$$CF_2Cl^{\cdot} + CFCl_2^{\cdot} \rightarrow C_2F_3Cl_3 \quad (8)$$

$$C_2F_2Cl_3^{\cdot} + F^{\cdot} \rightarrow C_2F_3Cl_3 \quad (9)$$

begünstigt sind, während die übrigen Reaktionen am langsamsten und gleichrangig sind.

Es zeigte sich, daß Veränderungen der Wertegruppen, aber auch von Einzelreaktionen den Kurvenverlauf stark beeinflußten und die Anpassungen an die experimentellen Werte signifikant verschlechterten.

Auch die ausgewählten günstigen Parameterwerte lassen auffällige Abweichungen und damit Merkmale des Systems erkennen. So liegen die Fluorid- und die Chloridkonzentration in reinem R113 mit $5 \cdot 10^{-4}$ bzw. 0,05 mol/l nahe am berechneten Kurvenanfangspunkt. Bei Arsenzugaben um 0,075 g erreichen die Ausbeuten jedoch höhere Werte als zu erwarten. Die Aufschlämmung kleiner Mengen Arsen könnte eine andere Oberflächenexposition ergeben, als sie bei großen Mengen auftritt. Die gemessenen Konzentrationen an $C_4F_6Cl_4$ scheinen im Bereich von 0,25 bis 1 g Arsen dem erwarteten Kurvenverlauf nicht zu entsprechen. Es ist hier eine systematische Abweichung zu vermuten. Sie könnte in Verbindung stehen mit einem anderen Ergebnis der Anpassungstests.

Es erwies sich als unmöglich, eine annähernd gleichmäßig gute Annäherung der Kurven an die experimentellen Werte zu finden, wenn nicht mit folgenden Werten der Diffusionskoeffizienten gerechnet wurde:

$$D_F = 1{,}5 \cdot 10^{-5}\ cm^2/s$$

$$D_{Cl} = 3{,}5 \cdot 10^{-5}\ cm^2/s$$

Dieses Resultat widerspricht der Erwartung, wonach das Fluorradikal schneller als das Chlorradikal diffundieren sollte. Es wurde damit die Vermutung nahegelegt, daß die Fluoratome rasch mit dem sich bildenden Chlor reagieren:

$$F^{\cdot} + Cl_2 \rightarrow ClF + Cl^{\cdot} \quad (10)$$

In diesem Fall würde sich neben den Chloratomen das ClF als das mit dem Arsen reagierende Produkt erweisen. Es wäre dann eine minimale stationäre Konzentration an Fluoratomen vorhanden und die Veränderung der Radikalkonzentrationsverhältnisse untereinander würde bei verschiedenen Arsenzusatzmengen auch die Bildung von R316 und den anderen Produkten beeinflussen. Gegenwärtig kann infolge der noch ausstehenden quantitativen Analysen weiterer halogenorganischer Reaktionsprodukte und der noch großen Streubreite der Ausbeuten eine genauere Überprüfung des Modells und gegebenenfalls eine Modifizierung nicht vorgenommen werden.

Die bisherigen Ergebnisse orientierten jedoch über die Möglichkeiten zur Erfassung der strahlenchemischen Vorgänge in heterogenen Systemen, die im Vergleich zur plasmachemischen Reaktionsführung von Interesse sind:

In feinteiligen Aufschlämmungen bzw. Suspensionen kann ein von der Festkörperoberflächengröße abhängiger Radikal- und Zwischenproduktabfang herbeigeführt werden, der an den Endproduktausbeuten verfolgbar ist. Die Beschreibung des Reaktionsgeschehens kann unter Einbeziehung von diffusionsbestimmten Reaktionstermen und dem 1. *Fick*schen Gesetz erfolgen. Eine unmittelbare Reaktion der halogenorganischen Radikale mit der Feststoffoberfläche braucht nicht einbezogen zu werden.

Eingegangen am 26. 9. 1984

Literatur

[1] *T. Tominaga, T. Fukumizu, Y. Makide*, Radiochem. Radioanal. Letters **21** (1975) 211

[2] *T. Tominaga, T. Aoyama, Y. Makide*, Radiochem. Radioanal. Letters **28** (1977) 27

[3] *U. Prösch, W. Uckert, L. Kolditz*, Isotopenpraxis **11** (1975) 63

[4] *U. Prösch, M. Luthardt, L. Kolditz*, Isotopenpraxis **13** (1977) 346

[5] *U. Westfahl*, Diplomarbeit, Humboldt-Univ. Berlin 1978

[6] *U. Prösch, K. Buttke*, Beiträge Plasmaphysik **24** (1984) 793

[7] *G. Paulin, E. Griepentrog*, Numerische Verfahren der Programmiertechnik. Berlin: VEB Verlag Technik 1975, S. 271

Isotopenpraxis 21 (1985) 8, S. 277 bis 281

Gamma-Radiolysis of Aqueous Deaerated Solutions of p-Hydroquinone

A. A. Al-Suhybani[1]), *G. Hughes*[2])

([1]) Dept. of Chemistry, King Saud University, Riyadh; [2]) Dept. of Chemistry, Liverpool University, Liverpool, U.K.)

The γ-radiolysis of p-hydroquinone (H_2Q) in deaerated solutions with different solutes is reported. At high pH there is no change in the system. When pH is lowered, H_2Q is consumed and $G(-H_2Q)$ is pH and $[H_2Q]$ dependent. Small yields of quinone (Q) were observed but $G(Q) \sim 10\%$ of H_2Q disappeared. $G(-H_2Q)$ is the same in different acids while $G(Q)$ and $G(H_2O_2)$ depend on the acid used. Competition studies were employed to evaluate the rate coefficient for the reaction of H with H_2Q. The values obtained support the reaction mechanism propsed. When N_2O is added, Q and 2-hydroxybenzoquinone (2-Q-OH) are produced, and their yields are pH dependent. A reaction mechanism is proposed to explain the experimental observations.

Über die Gamma-Radiolyse von p-Hydrochinon (H_2Q) in entgasten Lösungen mit unterschiedlichen Lösungsmitteln wird berichtet. Bei hohen pH-Werten tritt im System keine Veränderung ein. Wird der pH-Wert gesenkt, so wird H_2Q verbraucht und $G(-H_2Q)$ wird pH-Wert- und H_2Q-abhängig. Es wurde eine geringe Ausbeute an Chinon (Q) beobachtet, jedoch ist $G(Q) \sim 10\%$ H_2Q. $G(H_2Q)$ ist gleich bei verschiedenen Säuren, während $G(Q)$ und $G(H_2O_2)$ von der verwendeten Säure abhängt. Es wurden Vergleichsstudien durchgeführt, um den Geschwindigkeitsfaktor der Reaktion von H mit H_2Q zu ermitteln. Die erzielten Werte sprechen für den vorgeschlagenen Reaktionsmechanismus. Wird N_2O hinzugegeben, so entstehen Q und 2-Hydroxybenzochinon (2-Q-OH), deren Ausbeute vom pH-Wert abhängt. Zur Erklärung der experimentellen Beobachtungen wird ein Reaktionsmechanismus vorgeschlagen.

Keywords

G value; gamma radiation; kinetics; quinones; radiation chemistry; radiolysis

Introduction

The earliest work concerning the radiation chemistry of p-Hydroquinone (H_2Q) was performed by *Loiselear* [1] who used X-rays to irradiate dilute acid solutions of H_2Q in the presence of air and observed the formation of coloured oxidation products. When the irradiation was continued long enough, a reducing action finally took place and the colour became less. The work was later extended by others using γ- and X-irradiations under different conditions [2, 3]. They observed that the irradiation of H_2Q never yielded p-benzoquinone (Q). It was assumed that H_2Q was regenerated from H and the semiquinone radical ($H\dot{Q}$).

In the pulse radiolysis of H_2Q solutions, it has been shown [4, 5] that trihydroxycyclohexadienyl radical ($H_2\dot{Q}$-OH) is produced and its decay was found to be faster in O_2 solutions than in Ar solutions. This was explained by assuming, that in the presence of Ar, $H\dot{Q}$ disappears by the following reactions

$$H\dot{Q} + H\dot{Q} \rightarrow Q + H_2Q \quad (1)$$

$$H\dot{Q} + H_2\dot{Q}\text{-}H \rightarrow \text{Products} \quad (2)$$

where $H_2\dot{Q}$-H is the adduct of H with H_2Q. The present work is intended to test the above reaction mechanism, since if reaction (1) is occurring to a large extent in Ar solutions, Q would be a major product. In addition it would be of interest to see whether H_2Q is regenerated from H and $H\dot{Q}$, as assumed [2, 3], since if this is the case there will be no net change.

[1]) Address: P.O. Box 2455, Riyadh-11451, Saudi Arabia

Experimental and Materials

Triply distilled water was used in making all solutions. The irradiation of samples was carried out using a 90 Curie ^{137}Cs source at 20 ± 2 °C at dose rate 0·38 – 1·85 × 10^{19} eV L^{-1} min^{-1}. The irradiation facility has been described elsewhere [6]. Chemicals were analar grade where possible. Q was a BDH laboratory grade and was recrystallized twice from hot ethanolic solutions. 2-Hydroxy-p-benzoquinone (2-Q-OH) was prepared from 1,2,4-trihydroxybenzene by oxidation with Ag_2O in ethereal solution under strictly anhydrous conditions.

Q was determined spectrophotometrically using a (Unicam SP 500). The absorption maximum of Q is at 246 nm where its extinction coefficient (ε) was determined to be 2·20 × 10^4 M^{-1} cm^{-1}. At this wavelength ε for H_2Q was 186 M^{-1} cm^{-1} and is thus negligible. 2-Q-OH absorbs strongly at 246 nm (ε = 1·13 × 10^4 M^{-1} cm^{-1}). The amount of 2-Q-OH was determined from its peak at 365 nm (ε = 1·31 × 10^3 M^{-1} cm^{-1}). Its contribution to the absorption at 246 nm could therefore be calculated. H_2Q was determined from its absorption peak at 285·5 nm where ε was determined to be 2·50 × 10^3 M^{-1} cm^{-1}. It could also be determined by the reduction of ferriin and then measurement of the ferroin produced at λ_{max} = 510 nm. The two methods agreed quite well.

H_2O_2 could be determined either by the KI method [7] or by the Horseradish Peroxidase (HRP) method [8–10]. The latter method depends on the use of HRP as a catalyst for the oxidation of H_2Q to Q by H_2O_2 and hence measuring the absorption of Q. The first method was not practical because of the interference of H_2Q and/or Q. The second method is very accurate and it was checked against the KI method (in the absence of Q and H_2Q). In addition the method was tested by irradiating 1 M propan-2-ol/1 M H_2SO_4, O_2 saturated solution. The H_2O_2 produced was determined using the KI method and the HRP method and identical results were obtained.

Results and Discussion

In the radiolysis of 5 × 10^{-4} M H_2Q deaerated solutions at *p*H 5·8, no change was observed in the system, i.e. neither is Q produced nor does H_2Q disappear. When, however, the *p*H was lowered it was found that small amounts of Q were produced and the yield was [H_2Q] dependent. At the same time there was a significant consumption of H_2Q and this was *p*H and [H_2Q] dependent as can be seen in Tab. 1. G(-H_2Q) was also dose dependent as can be seen in Fig. 1. The G values are in all cases the initial yields. The anions of different acids react with OH with completely different rates. They follow the order Cl^- (at low *p*H)

Tab. 1. *G*(Q), *G*(H_2O_2) and *G*(−H_2Q) in the radiolysis H_2Q in deaerated solutions

Acid	10^4 [H_2Q] M	*G*(Q)	*G*(H_2O_2)	*G*(−H_2Q)
1 M H_2SO_4	20	0·30	0·61	3·49
1 M H_2SO_4	2	0·19	0·57	2·38
1 M H_2SO_4	1·9	0·22	–	2·38
0·05 M H_2SO_4	5	0·12	–	1·84
1 M HCl	2·2	0·47	0·21	2·53
1 M HCl	20	0·46	–	–
1 M HCl	10	0·45	0·14	–
0·1 M HCl	2	0·22	0·37	–
1 M $HClO_4$	2·4	0·11	0·78	2·36
1 M $HClO_4$	20	0·24	0·87	–

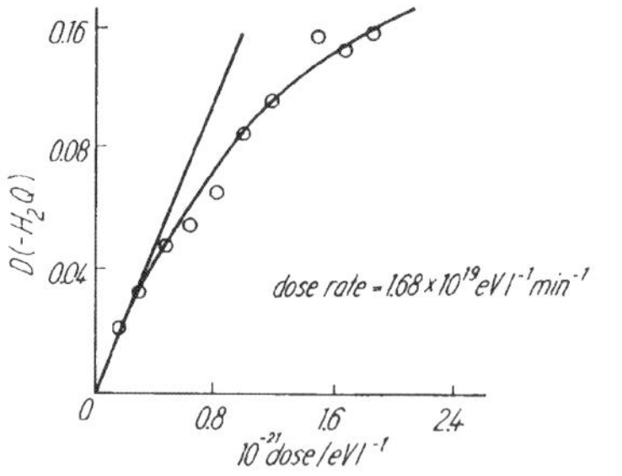

Fig. 1. Dependence of (-H_2Q) on dose in deaerated solution [H_2Q] = 17 · 10^{-4} M; [H_2SO_4] = 1 M

> HSO_4^- > ClO_4^- [11, 12]. The behaviour of these anions is reflected in the yield of molecular H_2O_2 · *G*(Q) was found to be independent of dose in HCl solutions while it depends on dose under other conditions.

In the radiolysis of water the following reactions explain the experimental observations:

$$H_2O \rightsquigarrow H_2O_2, H_2, OH, H_3O^+, e^-_{aq}, H, OH^- \quad (3)$$

$$e^-_{aq} + H^+ \rightarrow H \quad (4)$$

In deaerated water and in the presence of H_2Q, the following reactions will occur:

$$H_2Q + OH \rightarrow H_2\dot{Q} - OH \quad (5)$$

$$H_2\dot{Q} - OH \rightarrow H\dot{Q} + H_2O \quad (6)$$

$$H\dot{Q} \rightarrow \dot{\bar{Q}} + H^+ \quad (7)$$

Since the pK of $H\dot{Q}$ is 4 [13, 14], then most of the radical produced would be ionized and since no change is observed in the system under these conditions, then it would appear that the radical is reduced back to H_2Q by e^-_{aq},

$$\dot{\bar{Q}} + e^-_{aq} \rightarrow \bar{\bar{Q}} \quad (8)$$

$$\bar{\bar{Q}} + 2\,H^+ \rightarrow H_2Q \quad (9)$$

where $\bar{\bar{Q}}$ is, in fact, the doubly ionized H_2Q and reaction (9) follows (8) since pK_1 and pK_2 for H_2Q are 9·85 and 11·39 respectively [15].

Although it was possible to observe the esr spectrum of $\dot{\bar{Q}}$ when H_2Q is oxidized by passing its neutral solutions through a column of MnO_2 or by the air oxidation of H_2Q at *p*H 7, attempts to observe it by continuous irradiation of neutral solutions of H_2Q failed. The systems tried were 2 × 10^{-3} M H_2Q at pH 7, deaerated and air saturated, a mixture of 10^{-3} M H_2Q and 10^{-4} M Q (an ideal system for the formation of $\dot{\bar{Q}}$) and finally 5 × 10^{-4} M Q containing 1 M tert. BuOH at *p*H 8. These results indicate that there are some radiolytic species which react with $\dot{\bar{Q}}$ and not with its precursor since $\dot{\bar{Q}}$ has been observed by pulse radiolysis [13, 14]. This would support the occurrence of reaction (8).

The disappearance of H_2Q at low *p*H indicate that H produced in reaction (4) will react with H_2Q according to:

$$H + H_2Q \rightarrow H_2\dot{Q} - H \quad (10)$$

k_{10} = 1·3 × 10^9 M^{-1} s^{-1} [16]. An exclusive reaction between H and HQ similar to (8) is not apparentely occurring since this would lead to regeneration of H_2Q and no net change would be observed. The radical produced in reaction (10)

Energiewirtschaft:

UdSSR: KARPFENZUCHT MIT KKW-ABWAERME
In wenigen Monaten soll in der Sowjetunion ein erster Teil der Fischzuchtanstalt von Kurtschatow in Betrieb genommen werden und einen jaehrlichen Ertrag von 2000 Tonnen Karpfen auf den Markt bringen. Dieser Betrieb nutzt die Abwaerme des Kernkraftwerkes Kursk und ermoeglicht dank einer staendigen Wassertemperatur von 27 Grad die Zucht von Karpfen waehrend des ganzen Jahres. Zugleich ist vorgesehen, den anfallenden Kompost und das gereinigte Wasser der Fischzuchtanstalt einerseits zur Erzeugung von Champignons, andererseits zur Verwendung in Treibhaeusern heranzuziehen, in denen Fruechte, Gemuese und Blumen produziert werden. (H.P.E. nach Energie Nucléaire Magatine, Nr. 9, Febr. 1985) (SVA-Bulletin (1985) 7)

PERSPEKTIVEN DER FERNWAERME IN DER SCHWEIZ
Im Rahmen einer vom Schweizerischen Ingenieur- und Architekten-Verein (SIA) am 28./29.3.1985 in Bern durchgefuehrten Studientagung ueber Fernwaerme nahm der Praesident der eidgenoessischen Fachkommission fuer die Fernwaermeversorgung, Yves de Haller, unter anderem wie folgt zu den Moeglichkeiten und Grenzen des weiteren Ausbaus der Fernwaerme in der Schweiz Stellung:
Bis zum Jahre 2010 koennte durch Fernwaerme zwischen 12 und 15% der Waerme-Endnachfrage in den Gruppen Haushalt, Industrie und Dienstleistungen abgedeckt werden.
Als Anbieter von Fernwaerme wuerden in erster Linie die heute bestehenden und geplanten Kernkraftwerke auftreten. Hinzu kaeme ein wachsender Beitrag des Erdgases sowie der Abfallverbrennungsanlagen.
Da es sich bei der Waermeauskopplung in Kernkraftwerken um erprobte Technik handelt und keine besonderen Forschungs- und Entwicklungsarbeiten mehr erforderlich sind, kann in den kommenden Jahren ein bedeutener Schritt in der Substitution von Erdoelprodukten zurueckgelegt werden.
Aus den realistischen Szenarien ergibt sich, dass jaehrlich 1,5 bis 2 Mio t Erdoelprodukte substituiert werden koennten, und dies zu wettbewerbsfaehigen Preisen.
Die Kernenergie ist nach Auffassung von de Haller in besonderem Masse geeignet, einen ins Gewicht fallenden Beitrag zur Substitution und zur Diversifikation von Fernwaermequellen zu leisten. Fuer diese Beurteilung sprechen auch die in juengster Zeit in der Schweiz verwirklichten Projekte. (SVA-Bulletin (1985)8)

Hochenergiephysik:
CERN: PROTON-ANTIPROTON-KOLLISIONEN BEI 900 GeV
Am 17.3.1985 erreichten die Protonen- und Antiprotonenstrahlen im Super Proton Synchrotron (SPS) des Europaeischen Laboratoriums fuer Teilchenphysik (CERN) bei Genf erstmals die Rekordenergie von je 450 GeV. Diese extrem hohen Energien konnten dank einer neuen Technik, der "gepulsten" Beschleunigung, erzielt werden. Man hofft nun am CERN, durch die Beobachtung von Teilchenreaktionen bei Proton-Antiproton-Zusammenstoessen mit einer Gesamtenergie von 900 GeV erstmals im Labor seltsame und bisher unerklaerte Effekte studieren zu koennen, die aufgrund von Untersuchungen natuerlicher kosmischer Strahlen bei so hohen Energien zu erwarten sind. (SVA-Bulletin (1985)8)

Kernbrennstoffe:
NEUE GESELLSCHAFT FUER U-ANREICHERUNG UND ABFALLAGERUNG IN JAPAN
Am 1.3.1985 wurde die privatwirtschaftliche Unternehmung Japan Nuclear Fuel Industries Company, Incorporated (JNFI) gegruendet, deren Aufgabenbereich die Urananreicherung sowie die Lagerung schwachradioaktiver Abfaelle umfasst. Sie wird die kuenftige kommerzielle Zentrifugen-Anlage zur Urananreicherung mit einer Kapazitaet von 1500 Tonnen Trennarbeitseinheiten sowie ein Lager fuer 1 Mio 200-Liter-Faesser mit schwachradioaktiven Abfaellen errichten. Beide Anlagen gehoeren zum in der Praefektur Aomori geplanten Kernbrennstoffkomplex Rokkashomura. An der JNFI beteiligt sind die neun japanischen Elektrizitaetsgesellschaften, die Japan Atomic Power Company, die KKW-Lieferanten Hitachi, Toshiba und Mitsubishi Heavy Industries sowie verschiedene japanische Geldinstitute. (SVA-Bulletin (1985)8)

Kernenergetik:
ENTWICKLUNGSPERSPEKTIVEN IN DER UdSSR
In der UdSSR wird konsequent der Kurs verfolgt, eine grossangelegte, hocheffektive Kernenergetik zu schaffen, die in ihren technisch-oekonomischen Kennziffern einem Vergleich mit den traditionellen Ressourcen auf dem Gebiet der Elektroenergieerzeugung und in der Perspektive auch der Fernwaermeversorgung durchaus standhaelt. Derzeit gibt es moderne unifizierte Kraftwerksbloecke mit den Reaktoren WWER-1000 und RBMK-1000. Es wurde die Aufgabe gestellt, Bloecke mit einer Leistung von 1500 MW mit den Druckroehrenreaktoren RBMK-1500 und einer 750-MW-Turbine, einen Block fuer ein Kernheizwerk mit einem Reaktor mit einer thermischen Leistung von 430 Gcal/h (1797 GJ/h), einen industriellen Versuchsblock mit dem Reaktor BN-800 sowie einen Leitblock mit einem schnellen Reaktor vom Typ BN-1600 zu schaffen, um aus dem zur Verfuegung stehenden Kernbrennstoff groessere Mengen an Kernbrennstoff zu erbrueten. Im Verlauf der zweiten Etappe sollen dann energo-technologische Reaktoren und industrielle Versuchsanlagen fuer die gesteuerte thermonukleare Synthese geschaffen werden. Begonnen wird auch mit der Errichtung von Kernheizwerken und Kernheizkraftwerken fuer die Waermeversorgung von Grossstaedten sowie von Kernheizwerken fuer die industrielle Waermeversorgung (Heisswasser- und Dampferzeugung). Es steht die Aufgabe, den Bau schneller Reaktoren mit Brennstoffzyklusbetrieben zu beschleunigen, wodurch eine wesentliche Erweiterung der Produktion von Kernbrennstoff und eine Verbesserung des Einsatzes des Natururans moeglich werden. Es ist vorgesehen, in den naechsten Jahren die Wege zur Vervollkommnung der schnellen Reaktoren sowie zum praktischen Einsatz von Thorium und seinen Verbindungen im Kernbrennstoffzyklus zu untersuchen. Den Forschungs- und Entwicklungsorganisationen wurde die Aufgabe gestellt, die Probleme der Wiederaufarbeitung des in den Reaktoren verbrauchten Kernbrennstoffs und der Rueckfuehrung des unverbrannten Urans und Plutoniums in den Brennstoffzyklus zu loesen, schnelle Reaktorkomplexe mit einer erweiterten Kernbrennstoffreproduktion und kerntechnologische Komplexe zu schaffen, effektive Verfahren zur Demontage von stillgelegten Kernkraftwerken zu entwickeln, die ihren Normativbetriebszeitraum ueberschritten haben. (Energietechnik 35(1985)4, S. 121-127, S.M. Gortinski)

Strahlenschutz:
Leichte C-14-Kontamination von Arbeitern
Im kanadischen Kernkraftwerk Pickering wurde am 6.3.1985 festgestellt, dass sich mehrere Arbeiter, die fuer den Austausch von Druckrohren in den ersten zwei Bloecken beschaeftigt waren, mit C-14-haltigem Staub kontaminiert hatten. Der Nuklearkomplex Pickering liegt rund 30 km oestlich des Stadtzentrums von Toronto am Ontariosee. Er umfasst 8 Candu-Reaktorbloecke, von denen 7 im kommerziellen Betrieb stehen und deren achter demnaechst fertiggestellt wird. Die weiche Beta-Strahlung von C-14 wird bereits von einer drei Blatt dicken Papierschicht gestoppt. Weil sie so weich und wenig durchdringend ist, konnte mit den in Pickering routinemaessig eingesetzten Kontaminations-Detektoren zunaechst nichtsfestgestellt werden. Nach dem 6.3.. wurden alle Personen, die mit C-14 in Beruehrung gekommen sein koennten, auf eine moegliche Kontamination hin untersucht.

Dabei wurde bei 20 Untersuchten eine Kontamination der Haut festgestellt, und in 12 Faellen lag eine Verunreinigung der Kleidung, nicht aber der Haut vor. In allen Faellen lag die Hautdosis unter den vom Atomic Energy Control Board festgesetzten Dosisgrenzwerten. In den meisten Faellen lag die Dosis unter einem Prozent der zulaessigen Grenzwerte. Ebenfalls kontrolliert wurden die Wohnungen der kontraminierten Arbeiter. Dabei wurden in acht Wohnungen Spuren von C-14 festgestellt. Die gemessenen Strahlenpegel lagen so tief, dass keine Person einer Strahlendosis ausgesetzt war, die auch nur 1% der vom Atomic Energy Control Board fixierten Dosisgrenzwerte erreicht haetten. Waehrend einer einzigen Flugreise von Kanada nach Europa wuerde man eine groessere Strahlendosis erhalten. Der leicht radioaktive Leuchtstrumpf einer handelsueblichen Gas-Camping-leuchte haette in den verwendeten hochempfindlichen Detektoren viel hoehere Ausschlaege bewirkt als die verschleppte C-14-Kontamination. Um abzuklaeren, ob die Arbeiter C-14 inhaliert oder inkorporiert haben, wurden zunaechst routinemaessige Urinanalysen vorgenommen, die aber bei keinem Untersuchten Spuren von im Koerper aufgenommenem C-14 ergaben. Weitere Tests mit anderen Messtechniken sind noch im Gange. Bis zur vollstaendigen Klaerung der Situation wurden die Arbeiten zum Roehrenaustausch in den Einheiten Pickering-1 und -2 unterbrochen. Als Quelle der Kontamination wird ein schwarzes Pulver angegeben, das ausschliesslich im Block Pickering-1 zur Beschichtung von Druckrohren verwendet worden war. (SVA-Bulletin (1985)7)

Strahlenbiologie:
STRAHLENKREBS
Nach Einwirkung ionisierender Strahlen entstehende Tumore werden auf somatische Mutationen zurueckgefuehrt. Jetzt konnte erstmals eine konkret an der bei Maeusen durch Gammabestrahlung induzierten Karzinogenese beteiligte Punktmutation erfasst werden. Es fand sich eine Aktivierung des K-ras-Onkogens als Folge eines Basenaustausches in der DNA (Guanin-->Adenin) mit daraus resultierendem Aminosaeurenaustausch (Glycin-->Asparaginsaeure) in dem betreffenden Genprodukt.Nach Science 225(1984). (Deutsche Medizinische Wochenschrift 110(1985)239)

Lebensmittelbestrahlung:
LANDWIRTSCHAFTSPRODUKTE DURCH BESTRAHLUNG SCHUETZEN
In Marseille wurde mit den Vorarbeiten fuer das Projekt einer Anlage zur Bestrahlung landwirtschaftlich erzeugter Nahrungsmittel begonnen. Es ist vorgesehen, das Gesuch um Erteilung der entsprechenden Baubewilligung dem Service Cenral de Sûreté des Installations Nucléaires (SCSIN) im Juni 1985 zu unterbreiten und die Einrichtung gegen Ende des Jahres 1986 in Betrieb zu nehmen. Dabei sollen in zwei verschiedenen Abteilungen einerseits Lebensmittel-Halbfabrikate und deren Komponenten aus der Landwirtschaftsproduktion mit Gamma-Strahlen aus einer Kobaltquelle behandelt, anderseits Versuche zur Bestrahlung frischer Gemuese und Fruechte durchgefuehrt werden. Diese Methoden sind dazu bestimmt, die qualitative Beeintraechtigung der Lebensmittel durch Schaedlinge und Verunreinigungen zu verringern, solcherart also die Haltbarkeit der Produkte zu verbessern. Die neue Anlage in Marseille bedingt Investitionen in der Hoehe von FF 50 Mio. (SVA-Bulletin (1985) 8)

Japan:
LITHIUM AUS MEERWASSER
Im Hinblick auf die Gewinnung von Lithium aus Meerwasser hat das japanische Government Industrial Research Institute, Shikoku, ein mikroporoeses Adsorbermaterial aus Manganoxid entwickelt, das 5 bis 10mal wirksamer sein soll als herkoemmliche Adsorber. (SVA-Bulletin (1985) 8)

Autoradiography:
WHOLE-BODY AUTORADIOGRAPHY
Whole-body autoradiography is now being used rather extensively in toxicological research and screening. The purpose of the original whole-body autoradiographic technique was to map the distribution pattern of chemical compounds in experimental animals. This basic technique has later been further elaborated in various ways including the development of successively improved equipment for cryosectioning. It has also branched out in different directions, and some new technical approaches are discussed. The scope of whole-body autoradiography has increased by comprising also, e.g., volatile compounds ans short-lived isotopes (including double-tracer autoradiography). Recently a rapid autoradiographic technique has been worked out which allows work with the short-lived positron emitter C-11 (half-life 20 min.).[Int. J. Appl. Radiat. Isotopes 35)1984) 129, R.D.'Argy and S. Ullberg] Quantitation of an autoradiogram has been made more rapid and efficient by computerized image analysis. The technique has recently been used for studying not only distribution but also effect. The sectioning technique has also been developed for other purposes than just autoradiography, such as teratological screening, hard-tissue sectioning, a perfected way of dissection (taking samples for analysis with improved precision), and for detailed pathologic investigations of large human specimens. (Science Tools, 32(1985)1, S. 23, R.D'Argg and S. Ullberg)

Veröffentlichungen / Publications

NATIONAL ASSOC. FOR APPLICATION OF RADIOISOTOPES AND RADIATION IN INDUSTRY (NAARRI) BOMBAY,INDIA
NAARRI News 7 (1985) 1 Jan.-March
Annual General Meeting, Dec. 15, 1984

International Symposium on Artificial Radioactivity, Jan. 8-12, 1985,Madame Professor Helene Langevin-Joliot, daughter of Irene and Frederic Joliot and herself a well known nuclear physicist in France, delivered the Key note address on the "Historical Aspects of the Discovery of Artificial Radioactivity".

IAEA Interregional Training Course on Radiation Protection, Bombay, Dec. 12, 1984 - May 17, 1985

Workshop on "Radiopharmaceuticals and RIA Kits and their Medical Application" organised by the Bhabha Atomic Energy Centre (BARC), Bombay-400085, from Dec. 10 to Dec. 14, 1984.

There were a number of review talkes and discussion on various topics like nuclear cardiology, Tc-99m generators and formulations, cyclotron produced radiopharmaceuticals and PET, monoclonal antibodies, RIA of thyroid hormones and of infectious diseases etc.

Veranstaltungskalender – Tagungen – Lehrgänge / Calendar – Meetings – Courses

MODERN TRENDS IN ACTIVATION ANALYSIS CALL FOR PAPER
To commemorate the 50th anniversary of the first publication in the field of activation analysis by G. Hevesy and H. Levi of the Niels Bohr Institute in Denmark, the 7th International Conference on Modern Trends in Activation Analysis will be held in Copenhagen, 1986, June 23-27. This Conference deals with all aspects of contemporary activation analysis, and scientists from every field where nuclear activation analysis is contributing to scientific or technical progress are invited to attend.
Further information is available from
Dr. K. Heydorn
General Chairman MTAA-7
Riso National Laboratory, Postbos 49,
DK-4000 Roskilde/Denmark

13. WELTENERGIEKONFERENZ 1986 IN CANNES
Die 13. Weltenergiekonferenz, die vom 5. bis 11. 10. 1986 in Cannes stattfindet, steht unter

dem Leitthema "Energy: Needs/Expectations". Wie dem Vorprogramm zu entnehmen ist, wird dieser Sachbereich unter vier verschiedenen Aspekten beleuchtet, wodurch ein Bogen von den bisher gemachten Erfahrungen zu den in die Zukunft reichenden Erwartungen geschlagen werden soll. Die einzelnen Teile der Weltenergiekonferenz sind thematisch wie folgt gegliedert: Weltweite Energie-Entwicklungen seit 1970, Zusammenhaenge zwischen Wirtschaft, Umwelt und Energie, internationale Zusammenarbeit auf technischem und wirtschaftlichem Gebiet, kuenftige Zielsetzungen. Weiter Auskuenfte sind beim Sekretariat des Schweizerischen Nationalkomitees der Weltenergiekonferenz c/o Dr. Peter Stuerzinger, Elektrowatt AG, 8022 Zuerich, erhaeltlich. (SVA-Bulletin (1985) 7)

Ausgewählte Beiträge / Recently published papers

JADERNA ENERGIE 31(1985)4 (in Czech)

Application of natural zeolite tuff in cementation of radioactive concentrates - A. Pekár, J. Timulák

Neutron dosimetry with ionization chambers - F. Spurný, I. Votočková

Non-destructive methods for isotopic composition determination of spent fuel assemblies in reactors VVER 440 - P. Lipták et al.

Application of a microcomputer to the automation of the measurements with a microscope for examination of the nuclear emulsion - D. Blaho

IAEA-BULLETIN 26(1984)4

Introduction: Options for the long term - L. Konstantinov, H.-J. Laue

International status of HTGRs - J. Kupitz, J. Dee

Fast reactors - Updata of global developments - V. Efimenko et al.

Nuclear heat applications: World overview - J. Kupitz, M. Podest

CMEA plans for nuclear heat supply - A. Panasenkov et al.

SMPRs - Assessing global prospects - R. Schmidt

Survey of developments in nuclear fuel cycle - I. Rybalchenko, L. Konstantinov

Obninsk marks 30 years of nuclear power - A.M. Petros'yants

IAEA 28th General Conference: 1984 session highlights

Science - Two decades of progress at ICTP

Isotopes - Water research teams in Latin America - R. Gonfiantini

Nuclear standards - Status report from the IEC - J. Weill, M. Gandhi

IAEA-BULLETIN 27(1985)1

Overview: Perceptions, perspectives, and pro portions Safety considerations for spent fuel transport: View from the USA - R.M. Jefferson

Irradiation fuel transport - update from France - B. Lenail

International regulation: Clarifying roles - R.B. Pope

Physically protecting nuclear material - the legal framework - Ha-Vinh Phuong

Transporter worker exposures: How much in the UK? - Ken B. Shaw

Case history: Sinking of the Mont-Louis & nuclear safety - B. Augustin

Modern tools of the trade: Industrial radiation processing - V. Markovic

Nuclear plant economics: How competitive - L. Benett

Nuclear medicine in developing countries - M. Nofal

Radiopharmacy: New techniques spurring growth - H. V. Riuz

Managing for better nuclear plant performance - B.J. Csik

Dosimetry for quality control in radiation processing - J.W. Nam

Nuclear safety issues, codes, and guides - E. Iansiti

Reassesing the nuclear liability regime - Ha-Vinh Phuong

Nuclear power risks: Report from WHO - Dr.M.J. Suess

COLL. CZECHOSLOV. CHEM. COMMUN. 50(1985)1

Comparative study of cw CO_2 laser induced and SF_6 sensitized decomposition of CH_3J and CD_3J - J. Rejnet et al.

J. PHYSIC. CHEMISTRY 89(1985)2

Production of HO_2 within the track core in the heavy particle radiolysis of water - J.A. La Verne et al.

ATOMNAJA ENERGIE 59(1985)1 (in Russian)

Development of the Method of Surface Pseudosources for Calculating Neutron Distribution in a Cylindrical Reactor Cell - V.F. Boyarinov

Computer Simulation of Displacement Spikes in Irradiated Materials in the Continuum Approximation - V.P. Zhukov, A.V. Demidov

Beam Microwave Instability in Proton Synchrotrons - V.I. Balbekov, S.V. Ivanov

Calculation of Neutron Activation Effects of Nuclei for Signal Superposition in Photon-Activation Analysis - A.P. Ganzha et al.

Excitation Cross-Section of Characeristic X-Radiation by He-4 Protons and Ions for Elements in the Interval $22 \leq Z \leq 83$ - E. Brazevich

Effect of Reactor Irradiation on the Electrophysical Characteristics of p-Type Heteroepitaxial Silicon Films on Sapphire - B.V. Koba et al.

Influence Atmospheric Humidity Variations on the Distribution of Cosmic Background Neutron Fluxes Above Water Surfaces - E.M. Filippov

A Pyroelectric Gamma-Ray Detector with Compensation for Compton Electrons - V.A. Borisenok et al.

Polynomial Representation of Bremsstrahlung from Thick Targets for 10-22 MeV Electrons- V.E. Zhuchko, Zen Chan Uk

Coefficient of Heat Release for Oblique Flow Around Fuel Elements and Pipes - Yu.S. Yuryev, A.D. Efanov

The Application of Cadmium Telluride Detectors for Roentgenofluorescent Analysis of Uranium Solutions - V.V. Berdikov et al.

Spontaneous Fission Half-Lives of Pu-239 and Pu-241 - A.A. Druzhinin et al.

ATOMNAJA ENERGIJA 58(1985)6 (in Russian)
The Important Political Step - B.A. Semenov

Technical Principles of MAGATE Guarantees - D.L. Tolchenkov

Hydrogen Indicator for Leak Detection on "Sodium-Water" Steam Generators - F.A. Kozlov et al.

Computer Codes for Prediction of Thermohydraulic Parameters in Rod-Bundle Subassemblies - Yu.V. Mironov

An Evaluation of the Energy Dependence of the Average Number of Prompt Neutrons for Neutron-Induced Fission of Plutonium Isotopes - V.V. Malinovsky et al.

Controlling U-235 Content in Granulated Fuel - I.I. Kreindlin et al.

Migration of Radionuclides from Coolant into Stainless Steel Under BR-10 Reactor Conditions and a Deactivation Method - I.A. Efimov et al.

Preparation of Highly Active Fluoride Wastes for Storage by Melting - Yu.G. Lavrinovich et al.

On Contamination of Arctic Waters by Radioactive Effluents from Western Europe Nuclear Fuel Reprocessing Plants - S.M. Vakulovsky et al.

Spatial-Angular Distributions of Secondary Electrons for Gamma Irradiation - A.F. Adadurov et al.

Neutron Characteristics of Detectors with Hydrogenous Scatterers - M.P. Ivashkina et al.

Body Burdens of Artificial Radionuclides in Personnel of Research Reactor MP - A.A. Moiseyev et al.

Effects of Refined U-238 and Pu-239 Nuclear Data in Calculated Characteristics of a Fast Reactor Test Model - A.A. Vankov et al.

CZECHOSLOVAK J. PHYSICS 35B(1985)3
Proc. 3rd Internat. Symposium on "Recent Problems in Plasticity of Metals and Alloys" held at Mariánzké Lázné, 10 to 12 Sept. 1984

IZOTOPTECHNIKA 28(1985)1 (in Hungarian)
Principles of Nuclear Magnetic Resonance /NMR/ Imaging and Its Medical Diagnostic Adaption - M. Horváth

Silica Detector with Completely Emptied Layers - L. Sági, S. Deme

Radiohygienic Importance of Iodine Prophylaxis of the Population at Large - I. Turai et al.

Methodological Problems in the N-15 Isotope Dilution Method, II - J. Nosek, G. Bakonyi

Radiation Exposure from Incorporated Isotopes - F-né Beleznay

Radiation Exposure of Persons, Professionally Using Pharmaca Labelled with Radiotechnetium - Zs. Makra

Radiometers in Use in Hungarian Clinical Practice and Available on the Market, and Their Service - F. Farkas

NUKLEON (1985)1 (in Czech)
A Designer contra nuclear technology dessinateur-projeteur contre technologie nucleaire - J. Alexa

Comparison analysis of I-131-iodohippurate, II analyse comparative du I-131 iodohippurate, II - J. Moraves et al.

Allcut2 code for reliability evaluation of NPPs programme allcut2 pour evaluer la fiabilite des centrales nucleaires - J. Cyraniová

BULL. SOC. CHIMIQUE DE FRANCE (1984)11/12
Deuterium labelling study of the acid catalyzed cyclisation of methyl 19-novanticopalate - J. Bastard et al.

CHEM.-ING. TECHNIK 57(1985)1
Dekontamination von Anlagen zur Verarbeitung radioaktiver Leuchtstoffe - N. Hanebeck et al.

THE REV. SCIENT. INSTRUMENTS 55(1984)12
An automated cryogenic charcoal trap system for He isotope mass spectrometry - D.E. Lott, W.J. Jenkins

J. CHEM. PHYSICS 81(1984)12 Part 2
U isotopes effects in some ion exchange systems involving uranyl-carboxylate complexes - H. Young Kim et al.

J. CHEM. SOC., CHEM. COMM. (1985)2
Comformational dependence of the β D equilibrium isotope effect in acylic carbocations - U.-H. Siehl, H. Walter

PALAEOGEOGR.,PALAEOCHIM.,PALAEOECOL. 48(1984)2-4
Direct assessment of stable isotope variability in planktonic foraminifera populations - P. Schifferbein, S. Hills

GEOCHIM. COSMOCHIM. ACTA 49(1985)1
The stable carbon isotopes in enstatite chondrites and cumberland falls - P. Deiner, F.E. Wickmann

Alternatin of N-15/N-14 and C-13/C-12 ratios of plant matter during the initial stages of diagenesis studies utilizing archaeological specimens from Peru - M.J.D. Niro, C.A. Hastonf

High - O-18 granitic plutons from the frontenax axis, Grenville Province of Ontario, Canada - Y.N. Shieh

Chemical and isotopic evidence for mixing between depleted and enriched mantle, Northwestern USA - W.K. Hart

The D/H ratios of sap in trees - implications for water sources and tree: ring D/H ratios - J.W.C. White et al.

Anomalous isotopic compositions in the sea off Southern California - L. Brown et al.

J.CHEM. PHYSICS 81(1984)12 Pat 1
The high - resolution visible overtone spectrum of CH_4 and CD_3H at 77 K - G.J. Scherer et al.

Laser induced fluorescence of CH_3S and CD_3S radicals - M. Suzuki et al.

The lowest rotational transition of several isotopic forms of Kr D^+ - H.E. Warner et al.

The highly excited C-H stretching states of CH D_3, CH T_3, and CH_3 D - G.A. Voth et al.

Kinetics and mechanisms of the reactions of CH and CD with H_2 and D_2 - M.R.Berman,M.C.Lin

COMPT.REND.DER SEANC.SER.II,MECANIQUE,PHYSIQUE, CHIMIE 300(1985)1
Isotope effects in CO_2 formation during coal pyrolysis - R. Letolle et al.

SCRIPTA METALLURGICA 19(1985)1
A hydrogen peak of internal friction and its isotope effect in austenitic stainless steel - S. Asano, M. Kazaoka

J.AMER.CHEM.SOC. 107(1985)3
Chain growth rates in Fischer-Tropsch synthesis on an iron catalyst:An isotopic transient study - C.A. Mims, L.E. McCanadlish

may disappeare via,

$$H_2\dot{Q} - H + H_2\dot{Q} - OH \rightarrow \text{Products (s)} \quad (11)$$

$$H_2\dot{Q} - H + H\dot{Q} \rightarrow 2\,H_2Q \quad (12)$$

At high $[H_2Q]$, it may be assumed that all H react via (10). Under these conditions then

$$G(H_2\dot{Q} - H) = G_{e^-_{aq}} + G_H = 3{\cdot}6 \quad \text{and}$$

$$G(H\dot{Q}) = G_{OH} = 2{\cdot}9$$

If it is assumed that the rate coefficient of reactions (1), (2), (11) and (12) are about the same, then $G(-H_2Q) \sim 4$ and $G(Q) \sim 0{\cdot}8$. The predicted value of $G(\text{-}H_2Q)$ is close to the experimental, but $G(Q)_{calc.} > G(Q)_{meas.}$. This would indicate that $k_1 < k_2 + k_{11} + k_{12}$. It is possible that more Q is produced but removed by secondary reactions as it was found that $G(Q)$ was very much dependent on dose.

As mentioned earlier, the change of the anion has some effect on $G(Q)$ and $G(H_2O_2)$ as can be seen in Tab. 1. The effect of [HCl] on $G(Q)$ is explained by the following reactions occurring in the spurs:

$$H^+ + Cl^- + OH \rightarrow Cl + H_2O \quad (13)$$

$$Cl + Cl^- \rightleftharpoons Cl_2^- \quad (14)$$

Cl_2^- is an oxidizing agent like OH but since the overall concentration of the oxidizing species will not change by replacing OH by Cl_2^- and since the half-life of the later would be expected to be longer, there is a chance for it to escape the spur and hence reacts according to:

$$H_2Q + Cl_2^- \rightarrow H\dot{Q} + 2\,Cl^- \quad (15)$$

Decreasing [HCl] to 0·1 M reduces the extent of the spur scavenging and hence leads to an increase in $G(H_2O_2)$ which still $\sim 1/2\,G_{H_2O_2}$. This behaviour is in accordance with *Hayon*'s calculation that ~ 30% of $G_{H_2O_2}$ would be scavenged when the product of k[solute] is of the order of $10^7\,s^{-1}$ [17]. For reaction (13) in 0·1 M HCl, $k_{13}\,[Cl^-]$ is $\cong 4 \times 10^7\,s^{-1}$ and therefore Cl^- will scavenge some of molecular H_2O_2.

Effect of tert. BuOH and Propan-2-ol

tert. Butanol is one of few scavengers which has a wide difference in its reactivity towards OH and H, being much more reactive with OH, while propan-2-ol is more reactive towards H than tert. BuOH. The effect of these alcohols on $G(\text{-}H_2Q)$ is shown in Tab. 2.

Tab. 2. $G(-H_2Q)$ in the radiolysis of 2×10^{-4} M H_2Q in 1 M H_2SO_4 containing tert. BuOH and propan-2-ol, deaerated Solns.

[tert. BuOH] M	$G(-H_2Q)$	[propanol-2-ol] M	$G(-H_2Q)$
0	2·52	0	2·52
0·004	2·43	4×10^{-4}	2·75
0.012	2.70		
0·036	2·91	10^{-3}	2·52
0·072	3·15	4×10^{-3}	2·25
0·10	3·02	$1{\cdot}6 \times 10^{-2}$	1·10
0·20	2.38		
0·40	2·35	0·1	0
0·55	1.74		
0·75	1·38		
1·0	1·14		
1·5	1·02		

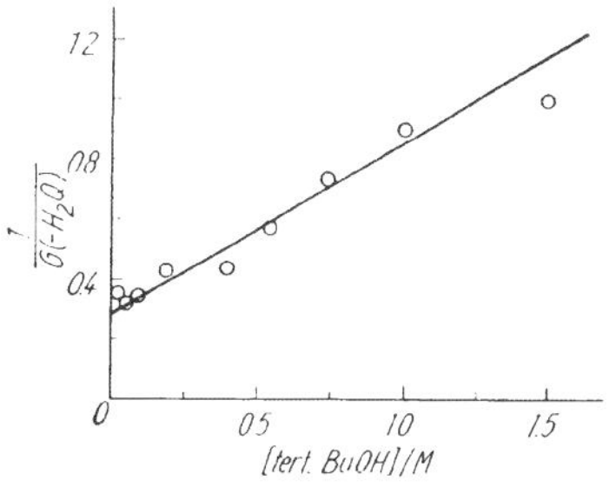

Fig. 2. Plot of $\frac{1}{G(\text{-}H_2Q)}$ V [tert. BuOH] in the radiolysis of 2×10^{-4} M H_2Q in 1 M H_2SO_4 deaerated solution

In the presence of tert. BuOH, in addition to reactions (5 and 10) the following reactions will occur,

$$(CH_3)_3COH + OH \rightarrow (CH_3)_2(\dot{C}H_2)COH + H_2O \quad (15)$$

$$(CH_3)_3COH + H \rightarrow (CH_3)_2(\dot{C}H_2)COH + H_2 \quad (16)$$

k_{15} is $5 \times 10^8\,M^{-1}\,s^{-1}$ [14] and k_5 is $5{\cdot}25 \times 10^9\,M^{-1}\,s^{-1}$ [4]. Thus at $[H_2Q] = 2 \times 10^{-4}$ M. < 5% of OH will react with H_2Q when [tert. BuOH] < 0·04 M. Under these conditions reaction (5) may be ignored and it can be shown that;

$$G(-H_2Q) = G(H)\,\frac{k_{10}\,[H_2Q]}{k_{10}\,[H_2O] + k_{16}\,[\text{tert. BuOH}]} \quad (17)$$

The above equation can be rearranged to:

$$\frac{1}{G(-H_2Q)} = \frac{1}{G(H)}\left(\frac{1 + k_{16}\,[\text{tert. BuOH}]}{k_{10}\,[H_2Q]}\right) \quad (18)$$

Plotting $\frac{1}{G(-H_2Q)}$ versus [BuOH] gives a straight line as can be seen in Fig. 2. Assuming $G(H) = 3{\cdot}6$ and $k_{16} = 10^5\,M^{-1}\,s^{-1}$ [14], it can be calculated from Fig. 2 that $k_{10} = 2{\cdot}4 \times 10^8\,M^{-1}\,s^{-1}$. This value is ~ 5 times less than reported for k_{10}, namely $1{\cdot}2 \times 10^9\,M^{-1}\,s^{-1}$ [16]. From the intercept of Fig. 2, $G(H)$ was found to be 3·5 compared with theoretical value of 3·6. These results support the assumption that there is no direct reaction between tert-butoxyl radical and H_2Q.

Propan-2-ol behaves in a similar manner to tert-BuOH though much lower concentration are needed to suppress $G(-H_2Q)$ as can be seen in Tab. 2. The reactions in this system are

$$(CH_3)_2CHOH + OH \rightarrow (CH_3)_2\dot{C}OH + H_2O \quad (19)$$

$$(CH_3)_2CHOH + H \rightarrow (CH_3)_2\dot{C}OH + H_2 \quad (20)$$

The fact that $G(\text{-}H_2Q) = 0$ at high [propanol-2-ol] rules out any reaction between $(CH_3)_2\dot{C}OH$ and H_2Q. A similar

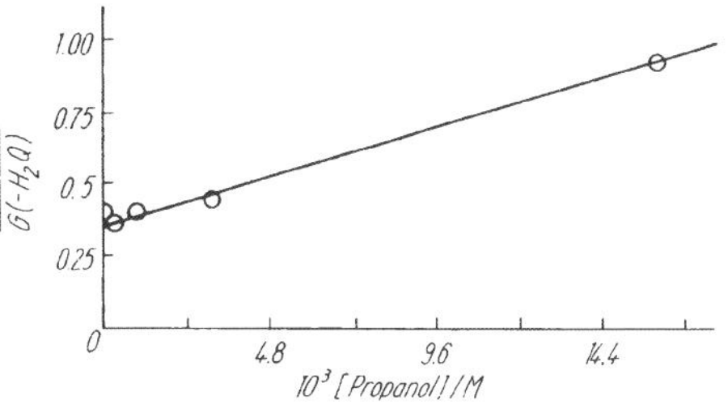

Fig. 3. Plot $\frac{1}{G(\text{-}H_2Q)}$ V [propan-2-ol] in the radiolysis of 2×10^{-4} M H_2Q in 1 M H_2SO_4 deaerated solution

expression to equation (18) can be obtained and leads to the plot shown in Fig. 3. Taking k_{20} as $5{\cdot}15 \times 10^7\ M^{-1}\,s^{-1}$ [18] and $G(H) = 3{\cdot}6$ gives k_{10} as $2 \times 10^9\ M^{-1}\,s^{-1}$. This value is very close to $1{\cdot}3 \times 10^9\ M^{-1}\,s^{-1}$ reported for reaction (10) [16]. The difference between k_{10} as determined using these two alcohols is probably due to the choice of the rate coefficient.

Effect of Nitrous Oxide

Q and 2-Q-OH are produced when 2×10^{-3} M H_2Q is irradiated in the presence of N_2O. Their yields are pH dependent as can be seen in Tab. 3. $G(Q)$ is independent of dose at $p\text{H} < 2{\cdot}6$ but decreases with dose thereafter. As would be expected there is very little Q at low pH where all e^-_{aq} are converted to H. $G(Q)_{meas}$, $G(Q)_{calc.}$ and $G(Q + 2 - Q - OH)_{meas}$ are plotted against pH as can be seen in Fig. 4 where it can be seen that a titration-type curves are obtained for $G(Q)_{calc}$ and $G(Q + 2 - Q - OH)_{meas}$. This may indicate that 2-Q-OH is produced from Q in a secondary reaction or it may indicate that both Q and 2-Q-OH have a common precursor. There is little effect of $[H_2Q]$ in the range $1{\cdot}7 \times 10^{-4} - 6 \times 10^{-3}$ M on $G(Q)$ and $G(2 - Q - OH)$ as can be seen in Tab. 4.

Tab. 3. Yields of Q and 2 – Q – OH in the radiolysis of 2×10^{-3} M H_2Q at different pH, in the presence of N_2O

pH	$G(Q)_{meas}$	$G(2-Q-OH)$	$G(\text{products})_{meas}$	$G(Q)_{calc}$
0	0·12	0	0·12	0
0·96	0·20	0	0·20	0
1·23	0·37	0·06	0·43	0·11
1·53	0·73	0·09	0·82	0·28
1·80	1·13	0·11	1·24	0·77
2·05	1·43	0·11	1·54	1·06
2·25	1·70	0·16	1·86	1·49
2·62	1·66	0·32	1·98	2·09
2·98	1·61	0·50	2·11	2·33
3·31	1·26	0·67	1·93	2·45
3·75	1·33	0·72	2·05	2·53
4·20	1·17	0·66	1·83	2·56

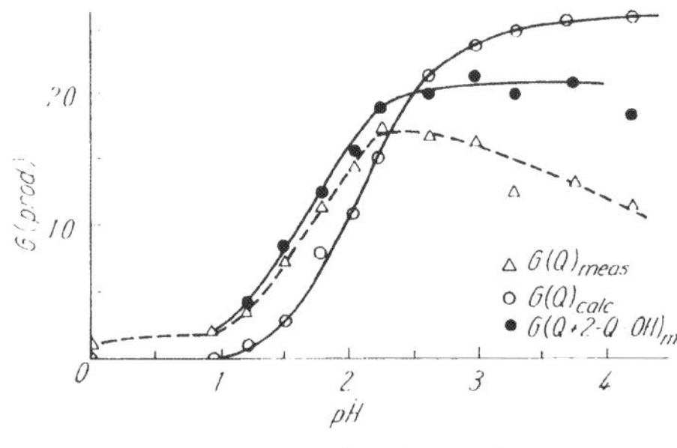

Fig. 4. $G(Q)_{meas}$, $G(Q)_{calc}$ and $G(Q + 2 - Q - OH)_{meas}$ as a function of pH in the radiolysis of 2×10^{-3} M H_2Q/H_2SO_4 solution, N_2O saturated

Tab. 4. Effect of $[H_2Q]$ on $G(Q)$ and $G(2 - Q - OH)$ in the radiolysis of H_2Q at pH 2·62, in the presence of N_2O

$10^3\,[H_2Q]$ M	$G(Q)$	$G(2-Q-OH)$	
0·17	1·36	0·40	$G(-H_2Q) = 1{\cdot}76$
0·4	1·46	0·36	
1·00	1·64	0·20	
2·0	1·66	0·32	
6·0	1·42	0·33	

In this system reactions (4, 5 and 10) are occurring in addition to:

$$N_2O + e^-_{aq} + H_2O \rightarrow OH + OH^- + N_2 \tag{21}$$

$$N_2O + H \rightarrow OH + N_2 \tag{22}$$

Reaction (22) is slow and can be neglected [19] H will react according to (10) where $G(H_2\dot{Q}\text{-}H)$ will depend on pH because of the competition for e^-_{aq} between H^+ and N_2O. In general $G(H\dot{Q}) > G(H_2\dot{Q}\text{-}H)$ and therefore all H_2Q-H will virtually react with $H\dot{Q}$. Therefore $G(Q)$ will be given by the following expression.

$$G(Q)_{calc} = 1/2\left[G_{OH} + \frac{k_{21}\,[N_2O]}{k_4\,[H^+] + k_{21}\,[N_2O]}\cdot G_{e^-_{aq}} - \left(G_H + \frac{k_4\,[H^+]}{k_4\,[H^+] + k_2\,[N_2O]}\cdot G_{e^-_{aq}}\right)\right] \tag{23}$$

$G(Q)_{calc}$ was calculated at different pH taking $G_{e^-_{aq}} = 2{\cdot}8$, $G_{OH} = 2{\cdot}9$, $G_H = 0{\cdot}55$ and $k_{21} = 8{\cdot}7 \times 10^9\ M^{-1}\,s^{-1}$ [20]. As can be seen in Fig. 4, $G(Q)_{calc} < G(\text{products})_{meas}$ upto pH 2·6 where the situation is reversed. The higher yield observed upto pH 2·6 is explained by spur scavenging by N_2O since $k_{21}\,[N_2O]$ is $17{\cdot}4 \times 10^7\ s^{-1}$. In addition the scavenging of e^-_{aq} in the spur would lead to N_2O^- which is a relatively long lived species and therefore may escape into the bulk of the solution [21] The lower values of $G(Q)$ at $p\text{H} > 2{\cdot}6$ are probably due to back reactions and this conclusion is substantiated by the fact that $G(\text{products}) = G(-H_2Q)$ and $G(Q)$ is dose dependent at this range of pH. Under the experimental conditions, most of e^-_{aq} will be scavenged by N_2O and therefore the most likely species to be responsible for these reactions are H and/or N_2O^-. The following reactions are suggested to explain the decrease in $G(Q)$ at high pH:

$$H_2\dot{Q} - OH + H^+ \rightleftharpoons (H_2Q - OH_2)^+ \tag{24}$$

$$(H_2\ddot{Q} - OH_2)^+ + N_2O^- \rightarrow H_2Q + H_2O + N_2O \tag{25}$$

$$H\dot{Q} + H_2\dot{Q} - H \rightarrow 2\,H_2Q \tag{26}$$

At high pH, production of 2 – Q – OH occurs with corresponding decrease in the yield of Q. Moreover $G(-H_2Q) = G$ (prod.). Earlier work in O_2-saturated solutions had indicated that 2 – Q – OH was produced as a result of combination of HO_2 and $H\dot{Q}$ but since there is no way in which HO_2 can be produced in this system, the formation of 2 – Q – OH is probably by the following reactions:

$$Q + N_2O^-(N_2OH) \rightarrow \dot{Q} - OH + N_2 \tag{27}$$

$$\dot{Q}\text{-}OH + H\dot{Q} \rightarrow 2 - Q - OH + H_2Q \tag{28}$$

where $\dot{Q} - OH$ is the adduct formed by the reaction of Q with OH radical.

Received August 1, 1984
Accepted in revised form January 2, 1985

References

[1] *J. Loiselear*, Bull. Soc. Chim. Biol. **25** (1943) 22
[2] *C. Vermeil, L. Salmon*, Comp. Rend. **249** (1959) 268
[3] *C. Vermeil, G. Roquet*, J. Chem. Phys. **60** (1963) 659
[4] *A. A. Al-Suhybani, G. Hughes*, J. Chem. Soc. Pak. (in press)
[5] *A. A. Al-Suhybani, G. Hughes*, Z. Phys. Chem. (in press)
[6] *G. Dobson, G. Hughes*, Trans Faraday Soc. **57** (1961) 1117
[7] *C. J. Hochanadel*, H. Phys. Chem. **56** (1952) 587
[8] *K. A. C. Elliott*, Biochem. J. **26** (1932) 1281

[9] *I. Yamazaki*, J. Biol. Chem. **235** (1960) 2444
[10] *J. Yamazaki, L. H. Piett*, J. Amer. Chem. Soc. **87** (1965) 986
[11] *C. J. Hochanadel*, Radiation Res. **17** (1962) 286
[12] *J. F. Ward, L. S. Myers*, Radiation Res. **26** (1965) 483
[13] *G. E. Adams, B. D. Michael*, Trans. Faraday Soc. **63** (1967) 1171
[14] *K. B. Patel, R. L. Willson*, J. Chem. Soc., Faraday Trans. I **69** (1973) 814
[15] *J. H. Baxendale, H. R. Hardy*, Trans. Faraday Soc. **49** (1953) 1140
[16] *O. I. Micic, M. T. Nenadovic*, J. Phys. Chem. **80** (1976) 940
[17] *E. Hayon*, Radiation Chemistry of Aqueous System. *G. Stein* (Ed.), New York: Wiley Interscience 1968, p. 157
[18] *M. Anbar, P. Neta*, Int. J. App. Radiat. Isotopes **18** (1967) 493
[19] *F. S. Dainton, P. Fowles*, Proc. Roy. Soc. (London) **A287** (1965) 295
[20] *J. Rabani, M. Pick, M. Simic*, J. Phys. Chem. **78** (1974) 1049
[21] *M. Anbar*, Adv. Chem. Ser. **50** (1965) 55

Isotopenpraxis **21** (1985) 8, S. 281 bis 284

Radiation Damage and Hydrogen Bonding in Biological Active Cyclohexane-1,2-diol and Pyranol in the Solid State

M. M. Abou-Sekkina, M. A. Sakran, A. Saafan

(Chem. Dept., Faculty of Science, Tanta Univ., Tanta, Egypt)

Biologically active 4-tert-butyl-1-methyl-3,3,4,5,5-pentadeuterocyclohexane-1c,2c-diol and cis-2-phenyltetrahydropyran-5-ol have been prepared in the solid state, recrystallized and identified. Extensive measurements have been carried out on these materials comprising infrared; electronic absorption spectra before, and after exposure to an absorbed energetic γ-ray dose (10^7 rd). These materials were chosen as they have intramolecular hydrogen bonding of variant origin. The relatvie strength of hydrogen bonding and magnitude of γ-radiation damage were evaluated in each case. Results obtained were in conformity, explained, assigned and discussed in details in relation with the nature and origin of hydrogen bonding in each case.

Biologisch aktives 4-tert.-Butyl--1-methyl-3,3,4,5,5-pentadeuterozyclohexan-1c,2c-diol und cis-2-Phenyltetrahydropyran-5-ol wurden in festem Zustand hergestellt, rekristallisiert und identifiziert. An diesen Materialien wurden umfangreiche Messungen einschließlich IR-Messungen durchgeführt; die elektronischen Absorptionsspektra wurden vor und nach der Bestrahlung mit einer energiereichen Gamma-Dosis von 10^7 rd ermittelt. Diese Verbindungen wurden deshalb gewählt, weil sie intermolekulare Wasserstoffbindungen verschiedenen Ursprungs aufweisen. In jedem Falle wurden die relative Stärke der Wasserstoffbindung und der Umfang des Gamma-Strahlungsschadens ausgewertet. Die erzielten Ergebnisse wurden in bezug auf die Natur und den Ursprung der Wasserstoffbindung erklärt, zugeordnet und diskutiert.

Keywords

binding energies; biological models; deuterium compounds; gamma radiation; radiation chemistry; radiolysis

Introduction

Although most reviews since 1920 (*Latimer* and *Rodebush* [1]) had the first definitive reference on hydrogen bonding.

Hydrogen bond is usually represented as A—H ... Y where A and Y are atoms with electronegativity greater than that of hydrogen. We will use the more general representation A—H ... B, where B is any σ or π electron donor-site (*Lewis* base) and A is one of the atoms (C, N, P, O, S, Se, Fe, F, Cl, Br, I).

Hydrogen bonds can be either intramolecular or intermolecular. For intramolecular hydrogen bonding to occur between A and B to form (A — B ... B) they must be in a favourable spatial configuration in the same molecule.

Two types of intermolecular hydrogen bonds, homointermolecular hydrogen bond (a) (self-association) refers to the association of two or more identical molecules, hetero-intermolecular hydrogen bond (b) refers to the association of two different species.

For the strength of hydrogen bonds, the reported enthalpy values for the reaction A—H + B—A—H ... B range from −0·5 kcal/mole for the thiophenol-benzene hydrogen bond [2] to 37 kcal/mole for the hydrogen bond n $(CH_3)_4\overset{+}{N}HF_2^-$.

Abou-Sekkina [3] investigated radiation damage and hydrogen bonding in some amino-anthraquinone dyes deducing a correlation between the extent of γ-ray damage and the energy of hydrogen bonding.

Thus, the major goal of the present manuscript is to evaluate the radiation damage of molecules containing intramolecular hydrogen bonding of different origins.

Experimental

Material Synthesis

1) 4-tert.-Butyl-1-methyl-3,3,4,5,5,-pentadeuterocyclohexane 1c,2c-diol [4]. To 4·72 g (30 mmol) 4-tert.-butyl-1-methyl-3,3,4,5,5-pentadeuterocyclohexene in 220 ml ethanol and 110 ml H_2O and 0·9 ml 10% NaOH, a solution of 12·64 g (80 mmol) $KMnO_4$ in 240 ml H_2O is added drop by drop with mechanical stirring at −20 °C. The solution was leaved to warm at room temperature. To this mixture a solution of 46 g $NaHSO_3$ in 120 ml H_2O was added. The solution was washed with ether 5 times and the total ether was also washed with 10% NaOH and water; dried over anhydrous Na_2SO_4.

The ether was evaporated and the material crystallised from petroleum ether (40 – 60 °C).

$C_{11}H_{17}D_5O_2$ (191·2): m.p. 82 – 84 °C; let. 80 – 81 °C; calculated: C 69·05 H + D 11·88; found: C 68·80 H + C 11·86

2) Cis- and trans-2-phenyltetrahydropyran-5-ol. To 1·7 g (45 mmol) sodiumhydride in 50 ml absolute dimethyleneglycoldimethyl ether (diglyme), 12·8 g (90 mmol) freshly distilled borontrifluoride in 20 ml absolute diglyme was added drop by drop at 23 °C under stirring. To this mixture, a solution of 9·5 g (60 mmol) 2-phenyl-3,4-dihydro-2H-pyron [7] in 50 ml absolute tetrahydrofuran was added [5 – 8].

After 90 min, the solution was warmed at 70 °C and the contents of the reaction vessel were cooled in ice bath with the addition of 10 ml 3 M NaOH and 10 ml 30% H_2O_2 drop by drop, which alone became very warm with stirring for 1 h. at 23 °C and washed with $CHCl_3$. The collected extract was then washed with water and dried over Na_2SO_4. The oil outstanding was distilled yielding 9·0 g (85%). 8·4 g from these cis, trans mixture dissolved in 25 ml acetone on kieselgel plates were well developed for 5 times using petroleum-ether (40 – 60 °C)/ether (1 : 1).

The two separated zones of the cis (slower) and the trans (faster) forms are present, the zones scratched in soxlet-appratus were extracted with 200 ml acetone. The solvent was then evaporated and recrystallized from petroleum ether (40 – 60 °C).

$C_{11}H_{14}O_2$: m.p. 79 – 80 °C; calculated: C 74·14 H 7·92; found: C 73·91 H 7·90

Infrared Absorption Spectra Measurements

The IR spectra of the materials investigated were carried-out at room temperature using the solid KBr disc technique and solution from material (0·005 M) in CCl_4. A Perkin Elmer (W. Germany) infrared spectrophotometer was employed in the range of 4000 – 200 cm⁻¹.

Electronic Absorption Spectra Measurements

The Nujol mull solid technique was followed for obtaining the room temperature electronic absorption spectra of the test samples.

A Unican SP 8000 Ultraviolet Recording Spectrophotometer was used in the range of 200 – 450 mμ.

γ-Radiation Exposure

The materials were exposed to an absorbed gamma dose of 10^7 rd using of γ-radiation from Egyptian Institute of Radiation Technology ^{60}Co gamma cell (activity 300 Ci), at room temperature, whereas the samples were exposed in air.

Results and Discussion

The appearance of γC = 0 (1700 cm⁻¹) in cyclohexanediol is an evidence for the oxidizing [10] action of the absorbed γ-dose (10^7 rd) on the secondary hydroxyl group (see Tab. 1).

This is most probably likely to be occurred through rupture of the cyclohexane ring.

Further support for this explanation, is the relatively pronounced red shift (towards Longer wave length) in the IR-spectral bands corresponding to γs-stretching vibration of (A – H) after irradiation or as a result of γ-radiation damage.

Tab. 1. Band Assignments of the Characteristic IR absorption bands

Material	M.P. (°C)	IR Absorption bands in cm⁻¹ Before irrad.	After irrad.	Band assignments ref. [9]
4-tert-Butyl-1-methyl-3,3,4,5,5-pentadeuterocyclohexane-1c,2c-diol	82 – 84	OH band		A – H
		3629	3620	γs-
		3585	3575	Stretching
		C = O	New band	
			1700	γC = O
Cis-2-phenyltetrahydropyran-5-ol	79 – 80	OH band		
		3602	3600	γs- A – H
		3630	3620	

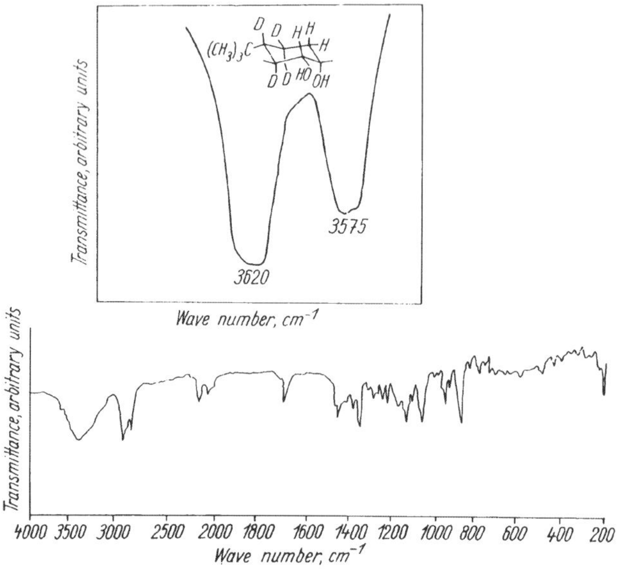

a

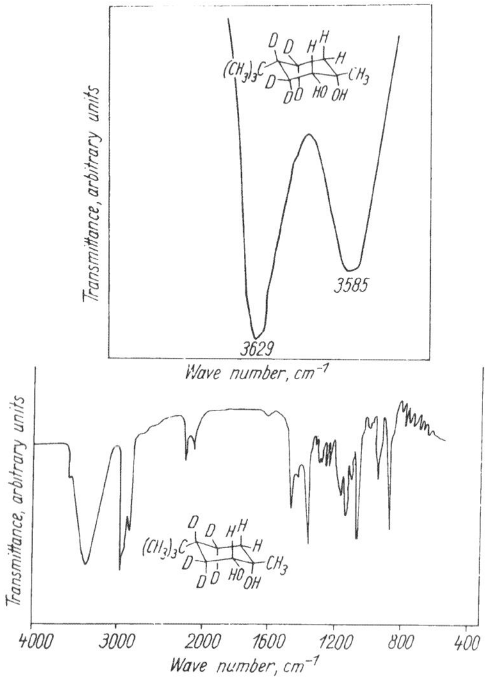

b

Fig. 1. The infrared absorption spectra of cyclohexandiol B) – before irradiation; A) – after irradiation (10^7 rd)

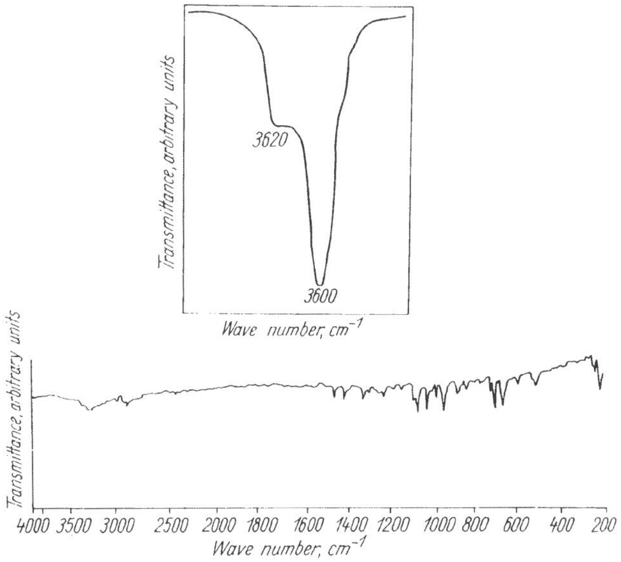

A

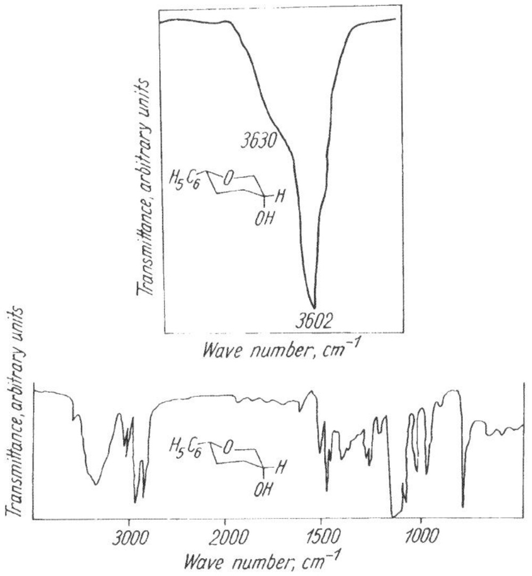

B

Fig. 2. The infrared absorption spectra of tetrahydropyranol
B) – before irradiation; A) – after irradiation (10^7 rd)

In cyclohexanediol, the intramolecular hydrogen bonding takes place between its two adjacent hydroxyl groups, whereas, in Pyranole, the intramolecular hydrogen bonding takes place between hydrogen of the hydroxyl group, and the ring oxygen.

Regarding the molecular structure formulae of the two materials investigated it can be seen that since C–O bond (in Pyranol molecule) is shorter [11] (1·43 Å) than the C–C bond (1·54 Å) in cyclohexane-1,2-diol, thus the intramolecular hydrogen bond in Pyranol would be stronger than that in cyclohexane-1,2-diol (see the chemical formulae).

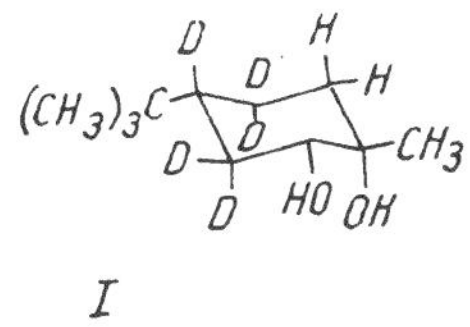

I

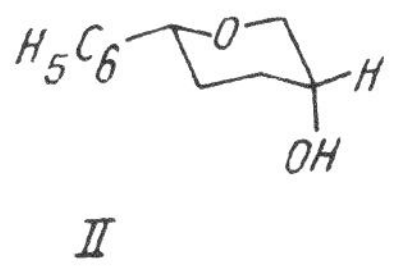

II

Tab. 2. Band assignments of the characterestic electronic absorption bands of materials investigated before and after the absorbed γ dose (10^7 rd)

Material	Electronic absorption		Band Assignments
	before irrad.	after irrad.	
4-tert.-butyl-1-methyl 3,3,4,5,5-pentadeutero-cyclohexane-1c,2c-diol	272	277	Electronic transition n – π*
cis-2-phenyltetradhydropyran-5-ol	332 273	335 279	

Fortunately, this explanation was further supported and established from our present results as given in Fig. 1 and 2, from these figures (1 and 2) it can be easily seen that why γ-induced IR shift is smaller in Pyranol than in cyclohexanediol.

As a result of the action of γ-radiation damage on electronic spectra of cyclohexanediol and pyranol, the following could be detected in common (see Tab. 2).

(1) The absorption bands are gradually shifted towards longer wave length (red shifted). This is probably correlated with the formation of decomposition centres vibrating at lower energy levels.

(2) The amplitude of some absorption bands are diminished after γ-ray absorbed dose and this is probably correlated with the γ-ray produced lattice imperfections which may lead to lattice defects and a decreased degree of crystallinity [10].

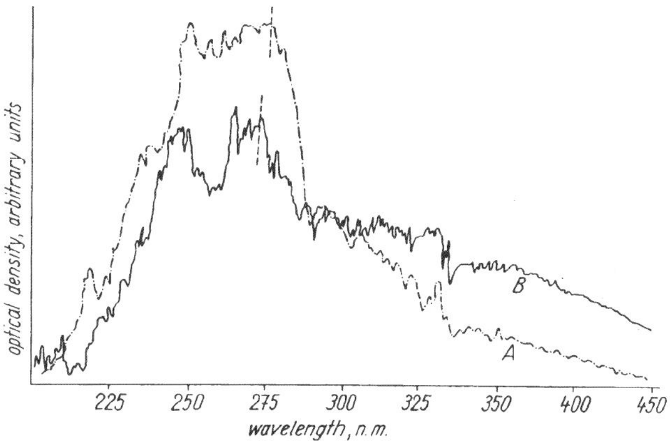

Fig. 3. The electronic absorption spectra of cyclohexanediol
B) – before irradiation; A) – after irradiation

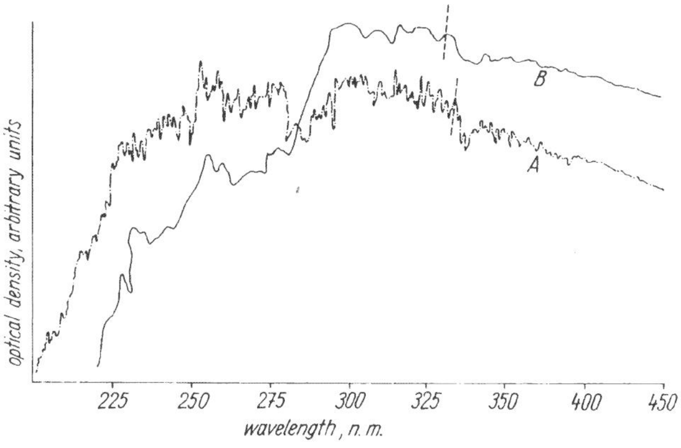

Fig. 4. The electronic absorption spectra of tetrahydropyranol
B) – before irradiation A) – after irradiation

(3) Some bands disappear and others are newly created and there is marked broadening and weakening of the absorption bands which is probably due to the decreased degree of crystallinity.

All of these changes, as a result of the absorbed γ-radiation dose (10^7 rd) damage, are pronouncedly occurred in cyclohexanediol. Again these could be explained on the same prementioned attributions as being differences in the nature of substituting groups and origin of hydrogen bonding in the two investigated materials.

Alternately results of IR and electronic absorption spectra were almost in conformity indicating that pyranol has greater resistance towards gamma absorbed dose damage (10^7 rd) than cyclohexanediol.

Received May 24, 1984

References

[1] *W. M. Latimer, W. H. Rodebusch*, JASC **92** (1920) 1419

[2] *R. Mathur, E. D. Becker, R. B. Bradley, N. C. Li*, J. Phys. Chem. **67** (1963) 2190

[3] *M. M. Abou-Sekkina, S. S. Assar*, Int. J. Appl. Rad. **32** (1981) 847

[4] *L. P. Barili, G. Bellucci, M. Macchia, F. Macchia, G. Parmigiani*, Gazz. Chim. Italiana **101** (1971) 300

[5] *N. H. Cromwell, R. E. Bamburg*, J. Org. Chem. **26** (1961) 997

[6] *N. L. Allinger, S. Greenberg*, JACS **84** (1962) 2394

[7] *E. M. Arnell, L. Joris, E. Milchell, T. S. S. R. Murly, T. M. Gorrie, P. V. R. Schleyer*, JACS **92** (1970) 2365

[8] *D. Hadzi, N. Kobilarov*, JACS (1966) 439

[9] *H. Ratajczak, W. J. Thomas-Orville*, J. Mol. Struct. **1** (1967) 449

[10] *M. A. Moharram, S. M. Rabie*, Isotope Rad. Res. **10** (1978) 91

[11] *B. J. Hendrickson, J. D. Cram, S. G. Hammond*, "Organic Chem." 3rd Edition. McGraw-Hill Kogakusha (1970). page 58

Isotopenpraxis 21 (1985) 8, S. 284 bis 286

Untersuchungen an Kieselglas mittels Positronenannihilation*)

G. Brauer, G. Boden

(Akademie der Wissenschaften der DDR, Zentralinstitut für Kernforschung Rossendorf)[1]

Mittels Positronenannihilation (Dopplerverbreiterung, Lebensdauer) können Aussagen zum Ordnungszustand verschiedener Kieselgläser gemacht werden. Kieselglas aus Quarz zeigt einen höheren Ordnungszustand als synthetisches Kieselglas aus $SiCl_4$; der Ordnungszustand von Kieselgläsern erhöht sich infolge Wärmebehandlung oberhalb 900 °C und hängt von der Herstellungstechnologie ab. Der Ordnungszustand wird im Rahmen des Vorordnungskonzeptes betrachtet.

Measurements of positron annihilation (Doppler broadening, lifetime) of different silica glasses led to statements about the degree of order of such glasses. Silica glass molten from quartz crystals showed a higher degree of order than synthetic silica glasses synthesized from $SiCl_4$; the degree of order of silica glasses increases at heat treatments above 900 °C and is influenced by the manufacturing technology. The degree of order is considered in connection with the pre-order conception.

Keywords

annihilation; Doppler broadening; fabrication; glass; heat treatments; order parameters; positrons; silicon oxides

1. Einleitung

Kieselglas, das wiederholt als einfache Modellsubstanz eines Glases herangezogen worden ist, besteht nach der Netzwerkhypothese aus einem ideal-ungeordneten SiO_4-Tetraedernetzwerk. Mit den verschiedensten physikalischen Methoden festgestellte „Heterogenitäten", d. h. Abweichungen von dieser Idealstruktur im sogenannten „röntgenamorphen Strukturbereich", sind in [1] zusammengetragen und vergleichend gegenübergestellt worden. Die zahlreichen Meßergebnisse aus der Literatur zeigen, daß sich durchaus graduelle Unterschiede des amorphen Zustandes in Abhängigkeit von der Vorgeschichte des Kieselglases, wie Schmelzvergangenheit oder nachträgliche Wärmebehandlung, erfassen lassen, ohne daß man jedoch exakte Strukturparameter, wie sie für kristalline Substanzen angebbar sind, ableiten konnte. Zusammenfassend lassen die aufgeführten Ergebnisse den Schluß zu, daß in einem aus Bergkristall erschmolzenen Kieselglas mikroheterogene Strukturanteile mehr oder weniger großen Ausmaßes vorhanden sind, deren Charakter mit dem von *Brückner* [2, 3] vorgeschlagenen Terminus „Vorordnung" recht gut beschrieben werden kann.

*) Vortrag, gehalten auf der 3. Tagung „Nukleare Analysenverfahren", Dresden/DDR 11. – 15. 4. 1983

[1]) Anschrift: 8051 Dresden, PSF 19

Neben den herkömmlichen Methoden zur physikalischen und chemischen Charakterisierung von Gläsern [4, 5] werden in zunehmendem Maße auch kernphysikalische Methoden zur Untersuchung der Struktur von Gläsern herangezogen. Vor allem wurde über Untersuchungen auf dem Gebiet der Tracerdiffusion [6] und der Kernreaktionen [7] sowie über Bestrahlungs-, [8] Lumineszenz- [9] und Streuexperimente [10] an Gläsern berichtet. Auch die Positronenannihilation (PA) findet auf dem Gebiet der Glasuntersuchung Anwendung, eine Methode, die seit dem Beginn der sechziger Jahre allgemein zur Untersuchung von elektronischen und strukturellen Eigenschaften von Festkörpern, Flüssigkeiten und Gasen mit herangezogen wird [11, 12].

2. Experimentelle Ergebnisse

2.1. *Bestimmung des Ordnungsgrades*

Es wird angenommen, daß sich amorphes und kristallines SiO_2 durch charakteristische Annihilationsparameter A_a bzw. A_c kennzeichnen lassen. Für ein Kieselglas mit unbekanntem Ordnungsgrad α, das sich durch den Annihilationsparameter A charakterisieren läßt, wird ein linearer Ansatz aufgestellt:

$$A = (1 - \alpha)\, A_a + \alpha A_c \qquad (1)$$

mit dem sich der Ordnungsgrad des Kieselglases in erster Näherung wiedergeben läßt. Für den Fall des röntgenamorphen Kieselglases kann α als ein der „Volumenkristallinität" vergleichbarer Ordnungszustand aufgefaßt werden. Diese Interpretation setzt eine Gleichverteilung der Positronen im Glas unter Vernachlässigung ihrer Diffusion voraus.

Als Standards wurden für ungeordnetes, nahezu idealamorphes Kieselglas (A, $\alpha = 0$) synthetisches, aus $SiCl_4$ in der Knallgasflamme hergestelltes Kieselglas und als vollständig kristalliner Quarz Brasilianischer Bergkristall (C, $\alpha = 1$) verwendet.

Der aus Dopplermessungen abgeleitete S-Parameter sowie die Intensität I_3 der langlebigen Komponente bei Positronenlebensdauermessungen wurden als Annihilationsparameter verwendet. Auf die Meßmethoden selbst sowie die Bestimmung von Annihilationsparametern wird ausführlich in [13] eingegangen.

Die Ergebnisse für ein aus Bergkristall hergestelltes Kieselglas (B, $\alpha = ?$) sind in Tab. 1 zusammengestellt. Daraus geht hervor, daß dem aus Bergkristall hergestellten Kieselglas ein höherer Ordnungsgrad als dem synthetischen Kieselglas zuzuordnen ist. Andererseits ist aus den Ergebnissen ersichtlich, daß beide Meßmethoden zum gleichen Resultat führen.

Tab. 1. S-Parameter, Intensität I_3 der langlebigen Lebensdauerkomponente sowie Ordnungsgrad α für verschiedene SiO_2-Spezies (A — synthetisches Kieselglas, B — Kieselglas aus Bergkristall, C — Brasilianischer Bergkristall)

Probe	S	I_3, %	α, %
A	0,3546 ± 0,0006	40,9 ± 0,6	0
B	0,3522 ± 0,0006		11,5 ± 1,4
		36,1 ± 0,5	11,7 ± 2,0
C	0,3338 ± 0,0006		100

2.2. *Einfluß der Temperbehandlung*

Der Einfluß einer 48stündigen Temperung auf den Ordnungsgrad von synthetischem Kieselglas im Temperaturbereich 900 °C bis 1400 °C wurde mittels Dopplermessungen untersucht. Die Ergebnisse sind in Abb. 1 dargestellt. Es zeigt sich, daß bei etwa 800 °C eine Zunahme von α gegenüber unbehandeltem Ausgangsmaterial ($\alpha = 0$) einsetzt, die mit steigender Temperatur stetig zunimmt. Der bei etwa 1250 °C beginnende starke Anstieg von α hängt mit der einsetzenden Oberflächenkristallisation zusammen,

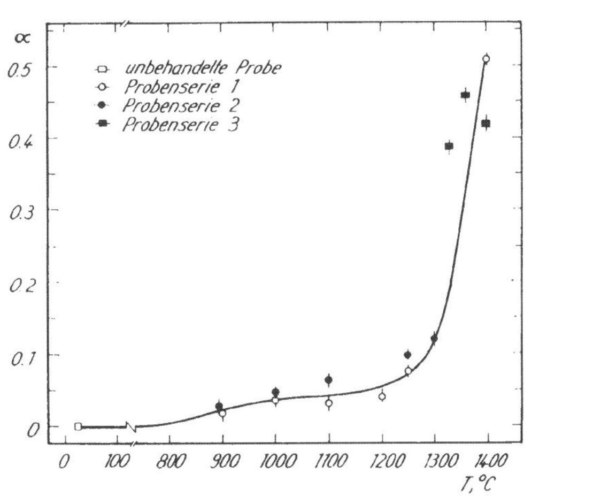

Abb. 1. Ordnungsgrad α von synthetischem Kieselglas nach Temperung (48 h) bei verschiedenen Temperaturen

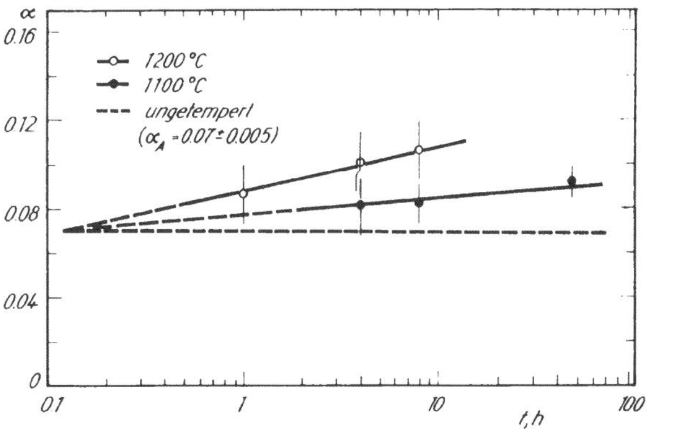

Abb. 2. Ordnungsgrad α von Kieselglas aus Bergkristall als Funktion der Temperzeit für verschiedene Temperaturen

die an den bei Temperaturen oberhalb 1300 °C behandelten Proben deutlich als weiße Schicht (Cristobalit) erkennbar ist. Die stetige Zunahme von α bei der Temperung ist letztlich als Vorstufe der von der Oberfläche ausgehenden Kristallisation anzusehen.

Um die Kinetik der Änderung von α näher beschreiben zu können, wurden Proben untersucht, die bei 1100 °C und 1200 °C verschieden lange getempert wurden. Als Ausgangsmaterial diente ein Kieselglas aus Bergkristall, dessen Ordnungsgrad vor dem Tempern $\alpha_A = (0{,}070 \pm 0{,}005)$ betrug. Die Meßergebnisse sind in Abb. 2 zusammengefaßt. Man erkennt, daß sich α mit zunehmender Temperzeit erhöht, wobei diese Erhöhung bei 1200 °C größer ist und auch früher im Experiment erkennbar ist als bei 1100 °C. Der Kristallisationsgrad wird von *Hautojärvi* u. a. [14, 15] durch eine *Johnson-Mehl-Avrami*-Gleichung beschrieben:

$$\alpha = \alpha_0[1 - \exp(-kt^m)] \qquad (2)$$

α_0 ist ein Sättigungswert, dem das System zustrebt. Mit k bezeichnet man eine Geschwindigkeitskonstante, die mit der Keimbildung, der Morphologie und der Kristallwachstumsrate zusammenhängt. Der Morphologieindex m dient als ein Maß für die Kristallisation in linearer ($m = 1$), flächenhafter ($m = 2$) oder räumlicher ($m = 3$) Dimension. t ist die Zeit, die die Probe dem äußeren Einfluß (z. B. einer Temperatur) ausgesetzt worden ist.

Es wird angenommen, daß sich die im Kieselgals stattfindenden Ordnungs- und Kristallisationsprozesse analog beschreiben lassen, wobei stets $\alpha_0 = 1$ gilt. Für Kieselglas aus Bergkristall muß Gl. (2) jedoch dahingehend modifiziert werden, daß der bereits im Ausgangszustand vorliegende Ordnungszustand α_A berücksichtigt wird:

$$\alpha = 1 - \exp(-kt^m) + \alpha_A/(1 + bt^c) \qquad (3)$$

Mit zunehmender Zeit nimmt gemäß dem in Gl. (3) willkürlich angenommenen Korrekturterm der Einfluß von α_A auf α entsprechend der Größe der Konstanten b und c ab. Anhand von Gl. (3) kann man unter der Bedingung $b = 0$ abschätzen, daß für die in Abb. 2 dargestellte Temperung von Kieselglas $m \leqq 0{,}4$ gilt.

An synthetischem, aus $SiCl_4$ in der Knallgasflamme hergestelltem Kieselglas wurde der Einfluß der Temperzeit auf α bei 1100 °C anhand von Dopplermessungen untersucht. Die Ergebnisse sind in Abb. 3 dargestellt und lassen sich gut durch Gl. (2) beschreiben. Das zeigt, daß sich hier ebenso wie in Kieselglas aus Bergkristall (Abb. 2) Ordnungsprozesse abspielen, die noch nicht die Bildung kristalliner Modifikationen ($1 \leqq m \leqq 3$) zur Folge haben,

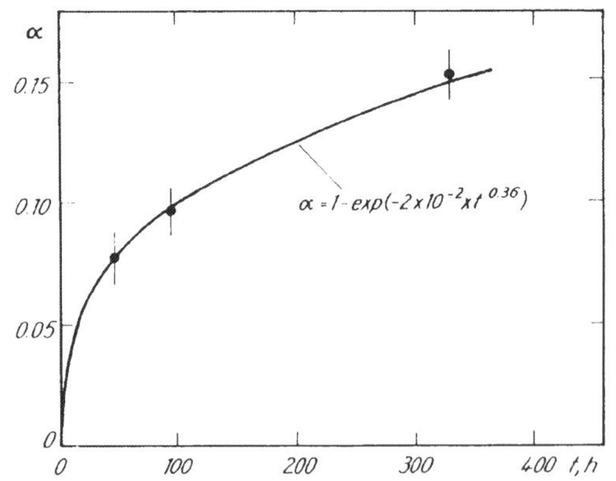

Abb. 3. Ordnungsgrad α von synthetischem Kieselglas als Funktion der Temperzeit bei $T = 1100$ °C

sondern auf einem niedrigeren Niveau — also beispielsweise als eine Erhöhung der Vorordnung im röntgenamorphen Bereich — ablaufen.

2.3. *Einfluß der Herstellungstechnologie*

Kieselglas aus Bergkristall wird durch Aufschmelzen von Naturquarz bei $T > 1750$ °C gewonnen. Da man zum Aufschmelzen verschiedene Technologien anwenden kann, ergab sich die Frage nach dem Einfluß des Herstellungsverfahrens auf α. Für die Untersuchungen standen Proben zur Verfügung, die alle aus dem gleichen Ausgangsmaterial nach vier verschiedenen Verfahren hergestellt worden waren. Die Ergebnisse sind in Tab. 2 zusammengestellt, wobei die gefundenen Unterschiede auf den Einfluß der Schmelzbedingungen zurückzuführen sind.

Tab. 2. Ordnungsgrad α für Kieselglasproben, die nach verschiedenen Verfahren hergestellt wurden (abs. Fehler von α: 1 %)

Proben-Nr.	Herstellungsverfahren	Schnittebene	α, %
1	Rohrschmelzanlage	radial	7,8
2	Rohrschmelzanlage	tangential	7,4
3	Rohrschmelzanlage	horizontal	8,2
4	Tiegelschmelze	radial	16,1
5	Tiegelschmelze	horizontal	16,2
6	Tiegelschmelze	horizontal	14,5
7	Tiegelschmelze	radial	16,5
14	Tiegelschmelze	radial	16,4
15	Tiegelschmelze	horizontal	15,8
16	Tiegelschmelze	radial	14,9
17	Tiegelschmelze	horizontal	15,2
8	Plasmaschmelze	radial	9,8
9	Plasmaschmelze	horizontal	9,1
10	Plasmaschmelze	tangential	9,7
		Schmelzlingsmitte	
18	Rotationsvakuumschmelze	tangential	7,0
19	Rotationsvakuumschmelze	radial	7,1
24	Rotationsvakuumschmelze	radial	7,8
25	Rotationsvakuumschmelze	tangential	7,5
		Schmelzlingsrandzone	
20	Rotationsvakuumschmelze	tangential	9,2
21	Rotationsvakuumschmelze	radial	8,9
22	Rotationsvakuumschmelze	radial	10,0
23	Rotationsvakuumschmelze	tangential	10,4

3. Schlußfolgerungen

Aus den dargelegten Ergebnissen geht hervor, daß mittels PA Aussagen zum Ordnungszustand von Kieselglas gemacht werden können. So wurde gezeigt, daß unter bestimmten Voraussetzungen und Annahmen das aus Bergkristall erschmolzene Kieselglas einen höheren Ordnungsgrad als das synthetische, aus $SiCl_4$ in der Knallgasflamme hergestellte Kieselglas aufweist, daß bei einer Temperbehandlung oberhalb 900 °C der Ordnungsgrad von Kieselgläsern zunimmt und daß die Herstellungstechnologie einen Einfluß auf den Ordnungsgrad des Kieselglases hat. Der Ordnungsgrad sollte dabei stets im Rahmen des „Vorordnungskonzeptes", wie es von *Brückner* [2, 3] aufgestellt worden ist, als ein Maß für noch im röntgenamorphen Strukturbereich befindliche, also noch nicht kristallisierte Strukturanteile betrachtet werden. Das Vorhandensein derartiger Strukturen konnte in der Vergangenheit durch eine Anzahl anderer physikalischer Untersuchungsmethoden gestützt werden, doch erweisen sich die meisten Methoden als nicht aussagekräftig genug, um für sich allein ausreichende Angaben zum Ordnungszustand zu machen.

Obgleich aus PA-Untersuchungen keine Angaben über definierte Bindungsabstände und -winkel zu erwarten sind — Aussagen darüber bleiben den Beugungsmethoden vorbehalten —, können vergleichende Werte über den Ordnungsgrad röntgenamorpher Festkörper einen Beitrag zum Verständnis des Aufbaus dieser Stoffe liefern. Dem zahlenmäßigen Wert des Ordnungsgrades kommt dabei zunächst nur insofern eine Bedeutung zu, als er im Rahmen der postulierten Randbedingungen ermittelt wurde und daher einen Näherungswert darstellt. Im Sinne von „erhöhter" bzw. „geringerer Ordnung" kann der Ordnungsgrad jedoch ein Hilfsmittel sein, um im röntgenamorphen Bereich den Ordnungszustand zu differenzieren.

Eingegangen am 26. 9. 1984

Literatur

[1] *G. Boden, W. Hinz*, Silikattechnik **29** (1978) 8

[2] *R. Brückner*, Glastechnische Berichte **37** (1964) 536

[3] *R. Brückner*, J. Non-Cryst. Solids **5** (1970) 123

[4] *H. Scholze*, Glas-Natur, Struktur und Eigenschaften. Braunschweig: Vieweg-Verlag 1965

[5] *W. Vogel*, Glaschemie. Leipzig: VEB Deutscher Verlag für Grundstoffindustrie 1979

[6] *G. H. Frischat*, Ionic Diffusion in Oxide Glasses. Aedermannsdorf: Trans. Tech. Publications 1975

[7] *A. Kolitsch, E. Richter, M. Wolf*, ZfK-361 (1978)

[8] *P. LeClerc*, Wirkung von Strahlen auf Gläser. B.I.S.T. Atomenergiekommission Nr. 98 (1965)

[9] *G. Boden*, ZfK-481 (1982)

[10] *J. Zarzycki*, J. Appl. Crystallogr. 7 (1974) 200

[11] *P. Hautojärvi, V. Komppa, I. Lehmusoksa, E. Pajanne*, J. Non-Cryst. Solids **18** (1975) 395

[12] *G. Brauer, G. Boden*, ZfK-442 (1981)

[13] *G. Brauer*, ZfK-483 (1982)

[14] *P. Hautojärvi, A. Vehanen, V. Komppa, E. Pajanne*, TKK-F-A326 (1977)

[15] *V. Komppa*, TKK-F-A351 (1978)

Isotopenpraxis 21 (1985) 8, S. 287 bis 290

Study of Filtration of Reactor Beam of Neutrons with Cadmium in a Multilayer Shield Containing Boron Carbide

R. M. Megahid

(Nuclear Research Center, A.E.E., Cairo, Egypt)

El-Sayed H. El-Kalla

(Physics Department El-Azhar University)

M. W. Esmaiel

(Faculty of Engineering Ain Shams University)

Experimental measurements have been carried out to study the effect of cadmium on the distribution and attenuation of reactor thermal neutrons emitted from a reactor core and the new thermal neutrons produced in a heterogeneous shield of water, iron, iron + B_4C and ordinary concrete. Measurements were performed using a reactor beam of neutrons filtered with cadmium emitted from one of the horizontal channels of ET-RR-1. It was found that the presence of cadmium sheet at channel exit causes a marked decrease in the thickness of the shield required to attenuate the thermal neutron flux by a certain factor.

Versuchsmessungen wurden durchgeführt, um die Wirkung von Kadmium auf die Verteilung und Schwächung thermischer Neutronen, die von der aktiven Zone eines Reaktors emittiert werden, sowie der in einer herterogenen Abschirmung, bestehend aus Wasser, Eisen, Eisen + B_4C und gewöhnlichem Beton, produzierten thermischen Neutronen zu untersuchen. Die Messungen wurden unter Verwendung eines aus einem der horizontalen Kanäle des ET-RR-1 austretenden kadmiumgefilterten Neutronenstrahls ausgeführt. Es wurde festgestellt, daß das Vorhandensein von Kamiumblech am Kanalausgang eine merkliche Reduzierung der für die Schwächung des thermischen Neutronenstromes um einen bestimmten Faktor erforderlichen Dicke der Abschirmung zuläßt.

Keywords

boron carbides; cadmium; flux density; gamma radiation; neutron flux; shielding materials; thermal insulation

Introduction

In a reactor shield heat may be generated, as 10% of the reactor power is carried away by the emitted radiation. The function of what so-called "Thermal Shield" is to reduce the energy flux of the emitted radiation to acceptable levels. Iron is frequently used as a thermal shield for nuclear reactors for its high density, high neutron capture cross-section, good thermal conductivity, besides it is cheap and easy to fabricate. A material with such properties is then able to absorb both core and capture gamma radiation; it concentrates much of the heat to a place from which it is easily conducted away [1—3].

The main problem in using iron, as a thermal shield, is due to the emission of high energy ($E = 7{\cdot}7$ MeV) secondary gamma radiation, as a result of radiative capture of thermal neutrons [3, 4]. This makes the most important source of gamma radiation, at the surface of the reactor shield. Consequently, iron has to be mixed with some materials, such as Boron carbide which has a high thermal neutron capture cross-section but which does not produce hard capture gamma radiation. This significantly reduces the contribution of the capture gamma radiation, to the total dose at the surface of the thermal shield. Thus the heat generated in the biological shield is reduced.

Few works [4—8] have been reported on the attenuation of neutrons and gamma radiation in laminated shields of different geometries. The present work is concerned with studying the effect of cadmium on the distributionend attenuation of reactor beam of neutrons in a heterogeneous shield of water, iron, iron + B_4C and ordinary concrete.

Experimental Details

The multilayer shields under investigation were arranged in front of one of the horizontal channels of the ET-RR-1 reactor. The shield medii were arranged on a large steel frame of 2·5 m length, 1·22 m width and 1 m height. The frame height could be shanged by means of a regulator fixed at the lower ends of its supports. At the front side of the frame there exists a steel plate, 0·7 cm thick and 35 cm in height, welded in the frame. This steel plate assures that the multilayer shield will be prevented from moving towards the reactor shield. The alignment of the shields with respect to the central line of the beam was carried out geometrically by adjusting the steel frame centre in coincidence with the centre of the beam.

In order to cut down neutrons scattered in the space between the multilayer shield and the reactor shield, a collimator, 70% by weight of paraffin and 30% by weight of boric acid, is used (see Fig. 1). The collimator is 40 cm length and has a central horizontal hole 14 cm in diameter.

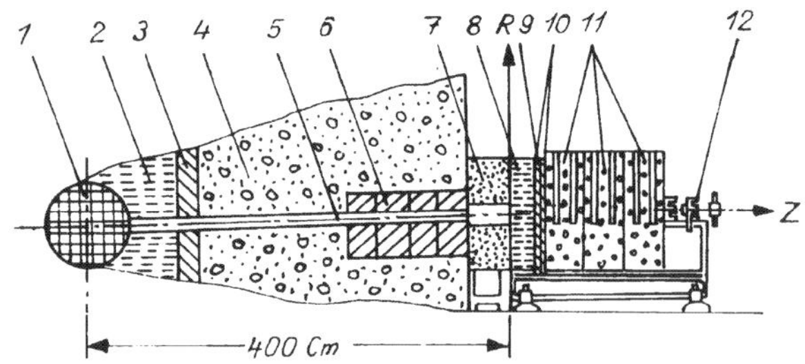

Fig. 1. The schematic diagram of the experimental layout
1 — Reactor core; 2 — Water cooling; 3 — Cast-Iron wall; 4 — Concrete wall of reactor; 5 — Experimental channel; 6 — Channel rotating gates; 7 — Collimator, 14 cm diameter; 8 — Water tank; 9 — Boron carbide filter; 10 — Iron slats; 11 — Investigated concrete blocks; 12 — Regulated for cassette height

This geometrical arrangement was quite satisfactory as it kept the background level around the experiment within the permissible occupational levels.

Thus, a laminated shield of water (100 × 100 × 20 cm³), iron (100 × 100 × 5 cm³) impregnated with boron carbide (1 cm) thick and ordinary concrete (120 × 120 × 120 cm³) was used during the present measurements [9]. The density of the ordinary concrete is 2·3 gm/cm³, where as the composition of concrete was given elsewhere [10].

Measurements were carried out along the beam direction (Z-direction) and normal to it (R-direction) using indium activation detectors. The activity was measured with an end window G-M tube.

The neutron detectors were fixed on special holders of density approximately equal to that of the medium [11]. These holders were inserted in both the water tank and concrete layer.

The uncertainty in both the irradiation time and reactor power variation was controlled by a phosphorus monitor fixed near to the reactor beam exit.

All sources of errors (due to uncertainty in sample position, sample weight, system efficiency and used cross-sections) were considered and estimated. In addition each set of measurements was repeated at least twice in order to reduce the effect of these uncertainties, especially for positions where the flux (or dose intensity) is low. However these sources of errors yielded an average error of 10%.

Results and Discussion

The total neutron fluxes of energies above 0·4 eV were obtained using a reactor beam filtered with cadmium.

The measurements of reactor thermal neutrons which emitted from reactor core were possible by substracting the data obtained using a neutron beam filtered with cadmium from that obtained using a direct beam.

The results may be classified into.

A. *Total neutron fluxes distribution using beam filtered with cadmium*

The attenuation curves for neutrons of energies initially above thermal energy in the laminated shield are represented in Fig. 2 and 3.

Fig. 2 shows that, the flux intensity of these neutrons decreased with the increase in the shield thickness for all values of R. This figure also shows that the rate of depression decreases with increasing R. It can be noticed from the curves that the decrease in flux intensity is smaller in water layer. However, the decrease shows a maximum

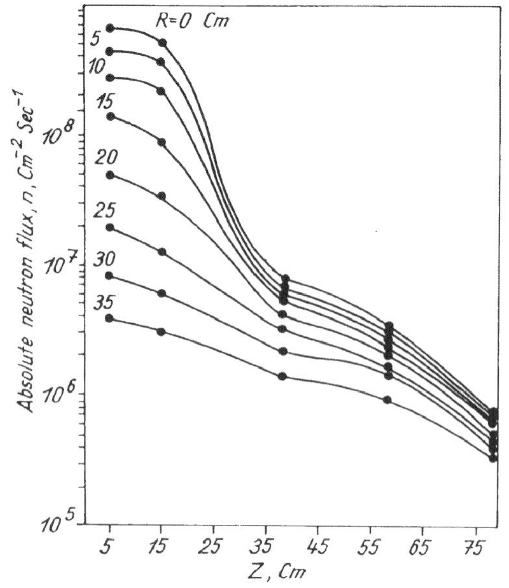

Fig. 2. Distribution of neutron fluxes, in laminated shield of water iron + B_4C and concrete in planes parallel to the beam axis, using a beam filtered with cadmium

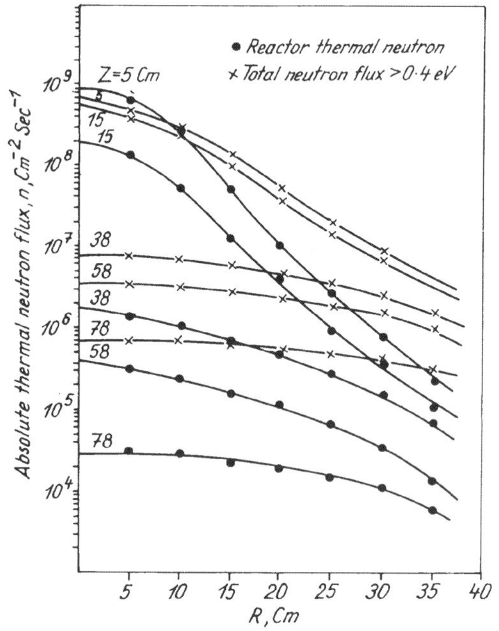

Fig. 3. Distribution of reactor thermal neutrons and neutrons (> 0·4 eV) fluxes in laminated shield of water, iron + B_4C and ordinary concrete in planes perpendicular to the beam axis

value at positions within iron layer and in the first layer of concrete shield. In case of water, the absorption of thermal neutrons by hydrogen nuclei is compensated with the newly produced thermal neutrons produced from the slowing down of neutrons of energies between 0·5 MeV and slow energies. The large decrease observed in flux intensity within iron and first concrete layer can be attributed to the high radiative capture cross section of iron and boron carbide for thermal neutrons. This capture process is not compensated with any another process which may result in the slowing down and thermalization of neutrons of higher energies. Within the concrete layer, the decrease in flux intensity is compensated to some extent with the new thermalized neutrons produced from the slowing down of neutrons of higher energies by conrete nucleii.

Fig. 3 shows that the rate of depression of $Z = 5$ and 15 cm is approximately the same. It can be noticed that the difference in flux intensity between the two curves is not too large. This is due to same reason mentioned before. These two curves also show that the rate of decrease has a maximum values at positions within the iron and the first concrete layer.

The attenuation curves constructed at planes located at $Z = 38$, 58 and 78 cm show that the decrease in flux intensity with increasing the concrete thickness is very small. This decrease may be due to the fact that the thermal neutron fluxes at these positions result mainly from the slowing down and thermalization of Scattered neutrons of higher energies. It can be also due to the divergence of the penetrating neutron beams.

Comparison between the group of isoflux curves which were taken from the attenuation curves represented in Fig. 2 and those for which were taken from the neutron distribution of a direct beam in the same shield [10] is shown in Fig. 4.

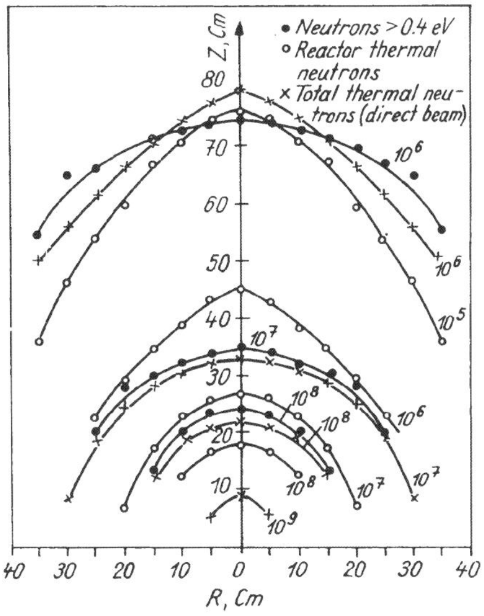

Fig. 4. Isoflux curves for neutrons > 0·4 eV, reactor and total thermal neutron fluxes in laminated shield of water, iron + B_4C and ordinary concrete using a beam filtered with cadmium

The results show that the presence of cadmium sheet at the channel exit causes a marked decrease in the thickness of the shield required to attenuate the thermal neutron flux by a certain value. It can be also noticed that the flux intensity is symmetrically distributed along the beam in the side directions.

B. *Distribution of reactor thermal neutrons*

Fig. 5 were constructed from Fig. 3. The results show that the reactor thermal neutron flux intensity decreases with increasing the thickness of laminated shield for all values of R. The rate of depression is found to decrease as R increases, and this is more noticable for values of $Z < 38$ cm. The attenuation of flux intensity with the increase in shield thickness is higher in water and iron regions. However, the decrease of total produced thermal neutrons has the smallest value in water layer as shown in Fig. 3. The decrease in flux intensity within water layer (especially for $R < 20$ cm) is attributed to the high absorption cross

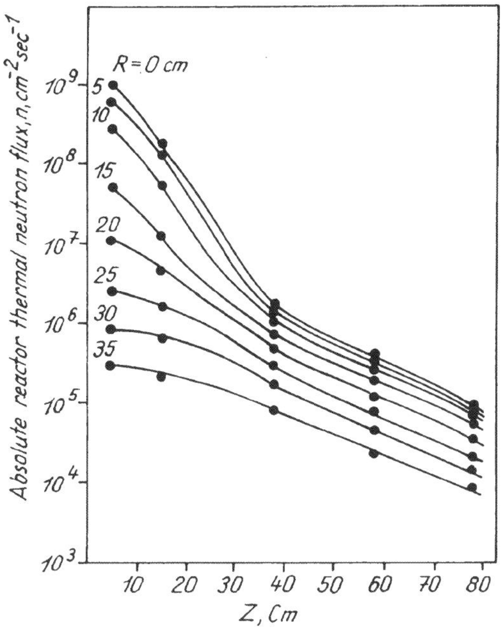

Fig. 5. Distribution of reactor thermal neutron fluxes in laminated shield of water, iron + B_4C and ordinary concrete in planes parallel to the beam axis

section of hydrogen for thermal neutrons. For $Z > 38$ cm and for all values of R, the flux intensity decreases exponentially with the increase in concrete thickness.

Fig. 3 shows also that for $Z = 5$ and 15 cm the attenuation of reactor thermal neutrons along the side direction is approximately the same and the rate of depression has a maximum value for $R > 5$ cm, i.e. outside the beam vicinity. It can be also noticed that the rate of depression in flux intensity of reactor thermal neutrons decreases with increasing Z. This may be attributed to the same reasons discussed before.

The isoflux curves for reactor thermal neutrons are represented in Fig. 4. Comparison between these isoflux curves and those obtained using a direct beam (total thermal neutrons) and cadmium filtered beams which are represented in the same figure show that the reactor thermal neutron emitted directly from the reactor core give the main contribution in the total measured thermal neutron flux at first positions. However, thermal neutrons produced from the slowing down of neutrons of energies above thermal give the main contribution at deep penetration.

Conclusions

1. The decrease in flux intensity due to the presence of cadmium filter is higher at positions along the beam axis and in planes with beam vicintiy. However, this effect is very small at positions far away from beam axis and entrance.

2. The attenuation of reactor thermal neutrons emitted directly from the reactor core in laminated shield show that the flux intensity decreases exponentially with increasing shield thickness through the three layers but, with a step at Z distances about 35 cm, it was found that the rate of depression in flux intensity is higher within the water and iron layers. However, the decrease in flux intensity of total and newly produced thermal neutrons has the smallest value in water layer.

3. Reactor thermal neutrons emitted directly from reactor core give the main contribution of total thermal neutron flux measured at first positions. However, at deep penetration new thermalized neutrons are the most predominant.

Received July 18, 1984

References

[1] *E. T. Price, C. C. Horton, K. T. Spinney*, "Radiation Shielding". Pergamon Press, Inc., New York 1957

[2] *H. Goldstein*, "Fundamental Aspects of Radiation Shielding". Pergamon Press, London, Paris 1959.

[3] *S. Glasstone*, "Principles of Nuclear Reactor Engineering". MacMellan and Co., Ltd., London 1955

[4] *V. D. Chitnis*, Hamma Ray Energy Absorption Rates in Iron and Lead. Dept. of Nucl. Eng., Iowa State Univ., U.S.A. 1968

[5] *T. Lefrert*, Nucl. Sc. Eng. **42** (1970) 267

[6] *R. Little, Valletri*, Trans. Am. Nucl. Coc. **1969**, 653

[7] *L. Harrisl, H. Kendrick*. Trans. Am. Nucl. Soc. **12** (1969) 959

[8] *H. Sprenger* et al., ORNI-T2-2658 (1972)

[9] *M. Abdel-Razek* et al., "The Effect of Cylindrical Air-Filled Ducts in Ordinary Concrete on the Distribution of Fast-Neutrons from a Collimated Beam". U.A.R.AEE-Rep. **146** (1967)

[10] *R. M. Megahid, El-Sayed H. El-Kalla, M. W. Esmaiel*, Egypt. J. Phys. **15** (1984) No. 1

[11] *R. M. Megahid, A. S. Makarious, M. A. El-Kolaly*, Ann. Nucl. Eng. **8** (1981) 79

Isotopenpraxis **21** (1985) 8, S. 290 bis 296

Buchbesprechungen

Grobstrukturprüfung mittels Röntgenstrahlung und Gammastrahlung

E. Becker. Leipzig: VEB Deutscher Verlag für Grundstoffindustrie 1983. 336 S., 288 Abb., 44 Tab.; Leinen M 48,— (Ausland M 52,—)

Die von einem erfahrenen Hochschulpädagogen und langjährigen Fachmann verfaßte Monographie ist eine „Einführung in die Theorie und Praxis der Durchstrahlungsprüfung von Werkstoffen", die beim Leser das Verständnis für den vielfach komplexen Zusammenhang zwischen Materialfehler, Prüfmethode, Bildqualität, Fehlererkennbarkeit und Bewertung wecken und die Prüfern in der Praxis sowie Studierenden an Fach- und Hochschulen die notwendigen Kenntnisse zur Durchführung von Grobstrukturuntersuchungen vermitteln soll.

Nach einem Abriß der Methoden und Apparaturen zur Strahlerzeugung und der Darlegung der Grundlagen der Durchstrahlung werden ausführlich die Nachweismethoden für Röntgen- und Gammastrahlung besprochen. Die Kapitel „Durchstrahlung mit Röntgen- und Gammastrahlung" und „Methoden zur Erweiterung des Objektumfangs" behandeln mit der gebotenen Gründlichkeit die Definition sowie internationale Standards zur Bestimmung und Kontrolle der Bildgüte eines Röntgenbildes sowie Mittel und Möglichkeiten zur Kontraststeigerung vor bzw. nach der Röntgenaufnahme; sind diese Verfahren doch die wesentliche Grundlage für eine zuverlässige Einschätzung der Fehlererkennbarkeit und somit der Qualität des geprüften Werkstückes. Mit der Technik zur „Verarbeitung industrieller Röntgenfilme" und der Zusammenstellung und Diskussion der „Filmbehandlungsfehler" wird in kompendiumartiger Weise ein Überblick über die Ursachen und Erscheinungsformen von Filmbehandlungsfehlern gegeben, deren solide Kenntnis eine Voraussetzung für die Vermeidung von Fehlinterpretationen von Durchstrahlungsaufnahmen ist. Abschließend werden Fehlertypen, ihre Ursachen, die speziellen Techniken zu ihrem Nachweis und ihre Erkennbarkeit im Röntgenbild (mit Beispielen) für die Haupteinsatzgebiete der Grobstrukturuntersuchung — die Prüfung von Schweißverbindungen, Gußstücken, Beton, Feuerfesterzeugnissen und elektronischen Bauteilen — behandelt.

Die Monographie empfiehlt sich durch ihre klare Gliederung und übersichtliche Darstellung dem Anfänger als Einführung, dem Fachmann zur Vertiefung; sie wird dank der zahlreichen tabellarischen Übersichten, dem Anhang „Röntgenbilder"-Beispiele sowie dem ausführlichen Sachwortverzeichnis darüber hinaus dem interessierten Leser als Nachschlagewerk willkommen sein. *H.-W. Thümmel*

Wissensspeicher Solartechnik
Thermische und fotoelektrische Nutzung der Solarenergie

H. Lippold, A. Trogisch, H. Friedrich. Leipzig: VEB Fachbuchverlag 1984. 324 S., 187 Abb., 79 Tab., 232 Lit.; Kunstleder M 30, —

Dieser informationsdichte Band setzt die Reihe der Wissensspeicher über moderne Gebiete der Technik des VEB Fachbuchverlag Leipzig in gleichartiger Ausstattung fort. Er gliedert sich in folgende Kapitel: 1. Überblick über die Nutzung der Solarenergie, 2. Grundgesetze der Solarstrahlung, 3. Aufbau und Berechnung von Solarkollektoren, 4. Solarkollektoranlage, 5. Beispielanlagen, 6. Fotovoltaische Wandlung der solaren Strahlungsenergie, 7. Solarzellen, Solarzellentechnologie und Entwicklungstendenzen, 8. Solarelektrische Systeme und Anwendungen; Literaturverzeichnis, Sachwortverzeichnis. Dem entsprechend vermittelt der Wissensspeicher neben physikalischem (und meteorologischem) Grundwissen den gegenwärtigen Stand der Technik bei der Niedertemperaturwärmeerzeugung in Solarkollektoren und der Energieerzeugung mit Solarzellen. Zahlreiche Abbildungen, Tabellen und Beispiele vermitteln anwendungsorientierte Fakten. Das Buch „soll nicht nur das Wissen um die ... Solartechnik erweitern, sondern ... auch helfen, Energiebewußtsein zu entwickeln" Dabei wird u. a. auf die in unseren Breiten begrenzten Nutzungsmöglichkeiten hingewiesen, die durch die relativ geringe Energiedichte und die starken Schwankungen im Angebot der Strahlung bedingt sind. Von der Freyburger Schwimmbadheizung abgesehen findet man über den Stand der Solartechnik in der DDR nur sporadische und unvollständige Mitteilungen. Dies wird der zugedachte breite Leserkreis zweifellos als Mangel empfinden. Davon abgesehen ist das Buch für alle empfehlenswert, die sich mit Solartechnik zu befassen haben. *K.-P. Dostal*

Die Beugung von Röntgenstrahlen und Neutronen in nichtidealen Kristallen

M. A. Kriwoglas. Kiew: Naukowo Dumka 1983. 407 S., 60 Abb., 9 Tab., 666 Lit. (in Russisch); geb. M 24,50

Die gemeinsame methodische Behandlung von Röntgen- und Neutronenstrahlung ist durch die größenordnungsmäßige Übereinstimmung der Wellenlängen charakteristischer Röntgenstrahlung und der *de-Broglie*-Wellenlänge thermischer Neutronen motiviert.

Der Autor beschreibt zunächst die allgemeine (harmonische) Analyse von Streuintensitätsverteilungen, ohne sich auf bestimmte Typen von Gitterdefekten festzulegen. Es schließt sich die Untersuchung statischer Verschiebungen in Kristallen mit begrenzten Defektbereichen an, und es folgen Ausführungen über Lage und Intensität von *Bragg*-Reflexionspeaks. Dabei werden sowohl Linienverschiebungen durch nichtideale Kristalle, Methoden zur Konzentrationsbestimmung von Zwischengitteratomen, Vakanzen usw., als auch der Einfluß von Wärmeschwingungen auf die *Bragg*-Intensitäten diskutiert. Unter der Überschrift „diffuse Streuung von Röntgenstrahlung und Neutronen durch Kristalldefekte" ist die Betrachtung nach zunehmender Kristallstörung geordnet. Schließlich wird die Röntgen- und Neutronenstreuung in Kristallen mit Versetzungen abgehandelt.

Das Buch hat rein theoretischen Charakter und ist auf Spezialisten für Beugungsmethoden zugeschnitten, also für einen sehr begrenzten Interessentenkreis gedacht. Die Lektüre ist nicht leicht, und alle Kapitel stehen in strengem, konsequentem Zusammenhang, wie sich u. a. aus zahlreichen Querverweisen ergibt. Ein Stichwortverzeichnis fehlt leider, was die Benutzung, gerade beim Nachschlagen, natürlich nicht erleichtert. Um einen einschlägigen Standardbegriff wie z. B. „rocking-Kurve" wenigstens dem Sinne nach herauszufinden, bedarf es relativ langwieriger Suche.

Es ist erstaunlich, daß sich ein Verlag unter solchen Voraussetzungen seitens des Manuskripts immerhin eine Auflage von 1300 Exemplaren zutraut, wenngleich natürlich in der SU hinsichtlich des Marktumfangs andere als DDR-Maßstäbe gelten.

Die zitierten Literaturstellen sind zwar zahlreich, meist aber alt.

G. Brunner

Cosmic Code
Quanten als Sprache der Natur

H. R. Pagels; aus dem Amerikan. übersetzt von R. Friese. Frankfurt/M – Berlin – Wien: Ullstein 1983. 352 S., 24 Abb.; geb. DM 38, –

Wieder einmal wird der so oft beanspruchte interessierte Laie angesprochen, sich mit Hilfe dieses Buches in der modernen Welt der Quanten und Quarks zurechtzufinden. Es ist vielleicht erwähnenswert, was den Anstoß zu diesem Buch gab: Im November 1977 fand and der Columbia-Universität ein Festkolloquium zu Ehren des Nobelpreisträgers *I. I. Rabi* statt. Wie so oft bei solchen Gelegenheiten, wurden brillante Reden gehalten, die zwar für den zu Ehrenden vorgetragen, aber im Grunde genommen an den versammelten Kreis gerichtet waren. Professor *Rabi* ging mit den Vortragenden ins Gericht; denn „sie hätten noch weniger als die Verfasser von Science-Fiction-Romanen dazu beigetragen, den Geist der Wissenschaft zu verbreiten ...", „... sie hätten es nicht fertiggebracht, der breiten Öffentlichkeit das Aufregende an der Physik nahezubringen ...". Und der Autor des Buches fährt fort: „Als ich ihn so reden hörte, beschloß ich, dieses Buch zu schreiben."

H. R. Pagels ist Professor für Theoretische Physik an der Rockefeller Universität und seit 1981 Präsident der New Yorker Akademie der Wissenschaften. Das Buch ist in drei große Abschnitte gegliedert: Der Weg zur Quantenrealität, die Reise in die Materie und der Kosmische Code. Eine Bibliographie und ein angemessenes Stichwortverzeichnis schließen es ab. Naturgemäß ist der erste Abschnitt auch der längste, werden doch dort die Grundsteine der modernen Physik, der Quantenphysik, gelegt. Statistische Mechanik, Unschärferelation, Relativitätstheorie werden anschaulich begründet und der Widerstreit der Meinungen dargestellt. Der zweite Teil ist den Elementarteilchen gewidmet, den Quarks und Leptonen, der Eichfeldtheorie und dem evtl. Protonengehalt und schließt mit einem kurzen Exkurs über die Entstehung des Weltalls ab. Der dritte und verhältnismäßig kurze Abschnitt stellt eine Art Zusammenfassung auf höherer Stufe dar, und so fragt der Autor: „Was ist das Universum? Ist es ein großer 3-D-Film, in dem wir alle unfreiwillig mitspielen? ... Das Universum ist für uns so schwer zu verstehen, weil wir es mit nichts vergleichen können... es ist eine in einem Code, einem kosmischen Code, abgefaßte Nachricht, und der Wissenschaftler hat die Aufgabe, diesen Code zu entschlüsseln."

Da das gesamte Buch ohne Inanspruchnahme der Formalismen der theoretischen Physik abgefaßt ist, dürfte der eingangs erwähnte naturwissenschaftlich interessierte Laie voll befriedigt sein; er wird viele Motivationen und Fragestellungen der Physiker besser verstehen und vielleicht Anteilnahme und Miterleben an den aufregenden Entdeckungen der Gegenwart empfinden, so bei dem noch mit Fragezeichen versehenen top-Quark in einer Tabelle des Buches. Das Buch ist sehr flüssig geschrieben, wie überhaupt die Übersetzung sehr gelungen ist, nur beim Inhaltsverzeichnis auf S. 9 fehlt die Angabe des 6. Punktes im Teil II. Man kann dankbar sein, daß sich der Autor durch die Hinweise von *I. I. Rabi* angesprochen fühlte und dieses Buch geschrieben hat.

H. Hübner

Ursprung und Zukunft des Weltalls
Pflanzen – Tiere – Menschen

J. Pfleiderer (Hrsg.). Mit Beiträgen von A. Fuchs, M. Gruber, L. Mandl, E. Reuer u. S. Reuer-Fabrizii. Frankfurt/Main: Umschau-Verlag 1983. 180 S., zahlr. S/W-Abb. und in Farbe; geb. DM 59,80

In einer Zeit, in der die einzelnen Wissensdisziplinen in immer speziellere Teilgebiete zerfallen, ist es ein schwieriges, aber auch sehr dankenswertes Unterfangen, eine Zusammenschau zu wagen und vorzulegen. Dieses Wagnis sind der Herausgeber – Prof. Dr. *Jörg Pfleiderer*, Ordinarius für Astronomie der Universität Innsbruck – und eine vergleichsweise geringe Zahl von Experten auf dem Gebiet der Astronomie, Botanik, Zoologie und Humanbiologie eingegangen. So erfährt der Leser und Betrachter des zahlreichen, wohlausgewählten Bildmaterials auf nur 48 Seiten zur Entstehung des Universums vom Urknall bis zu seiner derzeit überschaubaren Weiterentwicklung alles, was moderne Wissenschaft auf diesem Gebiet zu sagen hat. Es folgen die Entstehung des Lebens, die Entwicklung der Pflanzen und Tiere und die Entwicklung des Menschen.

Wer auch immer dieses Sachbuch zur Hand nimmt, der wird nicht nur ein verläßliches Wissen und neue Erkenntnisse aufnehmen, sondern auch einen bleibenden Eindruck von den vielen schönen Abbildungen haben, die ihn an die dargebotenen Zusammenhänge erinnern werden. Der weitere bleibende Eindruck dürfte die Erkenntnis sein, daß sich alles in der uns umgebenden Natur weiterentwickelt, seit Milliarden von Jahren und auch für künftige Milliarden von Jahren.

Welchen Stellenwert die geochronologischen Methoden mit Hilfe der radioaktiven Zerfallserscheinungen haben, ist deutlich hervorgehoben, denn sie erst erlauben eine Zuordnung der längst vergangenen Erdgeschichte in einer Zeitskala. Wie aktuell dieses Buch ist, kann an vielen Beispielen nachgeprüft werden: Vorstellungen über die Extinktion der Dinosaurier nach *Alvarez, Hsü* u. a., die Entdeckung der Quarks — wenn auch noch nicht des top-Quarks, was sich gegenwärtig noch in der Nachprüfung befindet —, die Auffindung von „Lucy", eines der ältesten Hominidenfunde im Hochland von Äthiopien, oder der Nachweis über die ersten Lebensformen in den Isua-Sedimenten von Westgrönland sind nur beispielhafte Hinweise.

Ein schönes, wohlausgestattetes Sachbuch, das den historisch-evolutionären Aspekt der Beschreibung der Natur überzeugend darstellt und zusammenführend weite Bereiche der Naturwissenschaft behandelt. Es wird von jedem, der sich über neue und neueste Entwicklungen in der Naturerkenntnis informieren will, dankbar aufgenommen werden, und jeder, der es nur anblättert, wird es dann ungern aus der Hand legen. *H. Hübner*

Amedeo Avogadro
A Scientific Biography

M. Morselli. Dordrecht — Boston — Lancaster: D. Reidel Publ. Comp. 1984, 375 S., 3 Abb., 8 Ill., 5 Tab.; geb. Dfl. 160,—; US $ 59.50

Amedeo Avogadro — oder genauer *Avogadro di Quaregna* — entstammt einer Adelsfamilie, die bis in die Mitte des 13. Jh. zurückverfolgt werden kann und deren Mitglieder zumeist als Juristen im Staatsdienst der Herrscher von Sardinien und Piemont standen, so auch sein Vater *Filippo Avogadro*. Amedeo hat deshalb selbst zunächst die Rechte studiert — er erwarb 1795 den Doktor-Titel im Kanonischen Recht — und stand danach bis 1806 im Staatsdienst. Bereits diese Feststellung ist wenig bekannt und bedarf der Beachtung; denn in fast allen Lebensbeschreibungen älteren und jüngeren Datums wird behauptet, er hätte Physik und Mathematik studiert. Dies hat er sozusagen in der Freizeit getan, er hörte Vorlesungen an der Universität von Turin — man nimmt an, seit ungefähr 1798. Er veröffentlichte 1806 eine kleine Arbeit über Elektrizität, verließ seine Stellung im Staatsdienst und wurde Dozent am Collegio delle Province und 1809 Professor für positive Philosophie (Mathematik und Physik) in Vercelli in der Nähe von Turin, ab 1820 dann an der Universität in Turin, wo er bis zu seinem Tode 1856 wirkte.

Das, was heute jedem Oberschüler in wenigen Zeilen einer Ableitung gemäß der kinetischen Gastheorie leicht verständlich gemacht werden kann, hat rund ein halbes Jahrhundert gebraucht, um sich endgültig durchzusetzen. Im Jahre 1811 veröffentlicht Avogadro sein berühmtes Essai und behauptet darin nicht nur, daß alle Gase bei gleichem Druck, gleichem Volumen und gleicher Temperatur die gleiche Anzahl Moleküle hätten, sondern auch daß die Moleküle aus zwei Atomen zusammengesetzt seien. Was aber fand Avogadro zu damaliger Zeit vor, um eine solche Behauptung aufzustellen? Es gab keine genauen Atomgewichte, man unterschied nicht zwischen Atom und Molekül (Avogadro selbst nennt diese zur Unterscheidung molécules élémentaires und molécules intégrantes), auch unterschied man nicht Atomgewicht und Äquivalentgewicht. Bekannt war lediglich das Gesetz der konstanten Proportionen von *Proust* (1797) und das der multiplen Proportionen von *Dalton* (1804), und *Gay-Lussac* verkündete 1808, daß die Gase untereinander im Verhältnis einfacher ganzer Zahlen reagieren. Allein die Annahme einer zweiatomigen Zusammensetzung der molekularen Gase rief einen heftigen Widerstand durch *Berzelius* hervor, der sein Lebenswerk in der Begründung eines elektrochemisch-dualistischen Systems sah. Selbst im Nachruf über Avogadro wird die berühmteste Arbeit aus dem Jahre 1811 nicht erwähnt. Den eigentlichen Durchbruch erfuhr diese Arbeit erst 1860 auf einem Chemiker-Kongreß in Karlsruhe, wo *Cannizzaro* eine von ihm veröffentlichte Schrift aus dem Jahre 1858 verteilte und damit einem breiten Kreis von Chemikern Avogadros Arbeiten bekannt machte. Nach Aussage von *Lothar Meyer*, einem Teilnehmer des Kongresses, hätte diese Schrift eine solche Klarheit gehabt, daß es ihm „wie Schuppen von den Augen fiel; die Zweifel schwanden, und das Gefühl ruhigster Sicherheit trat an ihre Stelle".

Das vorliegende Buch bietet in einer bewundernswerten Fleißarbeit Einblicke in eine Zeit, in der sich die Atomistik entwickelte und an deren Beginn Avogadro stand. Wir haben heute einen entwickelten Sinn, welche Richtung die Dinge bei ihrer Ableitung und Begründung nehmen sollen, aber dies war ein langer und mühsamer Prozeß, der in dem Buch dargestellt wird. Hervorzuheben ist die gelungene Verknüpfung von Avogadros Hypothese mit dem Verhalten und den Reaktionen aller großen Chemiker seiner Zeit. Wer auch immer dieses in souveräner Form erarbeitete Buch liest, wird zutiefst überzeugt sein, daß eine der Grundkonstanten der Natur zu Recht seinen Namen trägt — die *Avogadro*sche Konstante. *H. Hübner*

Radioactive Waste Management
Proc. of an International Conference, Seattle, USA, 16—20 May 1983

Vienna: IAEA 1984. Vol. 1—5. 2448 S., div. Abb. und Tab. (Proc. Series, CN — 43, STI/PUB/649); karton. ÖS 4650,—

An der Konferenz nahmen 528 Vertreter aus 29 Staaten und acht internationalen Organisationen teil. Es war die erste weltweite Konferenz zur allseitigen Betrachtung des Standes bei der Lösung der Fragen, die mit radioaktivem Abfall verbunden sind. Die 149 Vorträge der Konferenz galten folgenden Problemkreisen:

- Abfallpolitik und ihre Anwendung im nationalen und internationalen Maßstab (rechtliche, ökonomische, soziale und Umweltaspekte). (Vol. 1)
- Handhabung, Behandlung und Konditionierung von radioaktivem Abfall aller Kategorien einschließlich gasförmigem Abfall, Abfall aus Kernkraftwerken einschließlich deren Stillegung, aus Wiederaufarbeitungsanlagen und von Anwendern von Radionukliden. (Vol. 2)
- Lagerung und Endlagerung von radioaktivem Abfall aller Kategorien (nationale Konzeption, Projekte für konkrete Anwendungsfälle, Beschreibung von Untergrundlaboratorien, Erfahrungen mit oberflächennahen

Endlagern für niedrig- und mittelaktiven Abfall). (Vol. 3)

- Umweltbetrachtungen und Sicherheitseinschätzungen beim Umgang mit radioaktivem Abfall (Ziele und Methoden zur Einschätzung von Endlagerung in tiefen geologischen Formationen, von Endlagerung für niedrig- und mittelaktiven Abfall sowie von Abfällen aus Uranaufbereitungsanlagen. (Vol. 4)
- Abgabe von Radioaktivität in die Umwelt aus dem Betrieb von Kernanlagen (Stand und Entwicklung, Transportprozesse und deren Modellierung, Kontrolle von Endlagern auf dem Meeresboden). (Vol. 5)

Entsprechend dem sehr breiten Profil der Konferenz geben die Vorträge sowohl den in Planung und Überwachung als auch den in Wissenschaft und Produktion Tätigen einen guten Überblick. Die Materialien der Konferenz können ferner auch als Modell bei der Lösung von Problemen mit anderen Schadstoffen dienen.

Auf eine ausführliche Konferenzberichterstattung in der Zeitschrift Kernenergie **27** (1984) 265–275 sei abschließend hingewiesen.

K. Ebel

Laboratory Training Manual on Radioimmunoassay in Animal Reproduction

Vienna: IAEA 1984. 270 S. div. Abb. u. Tab. (Technical Reports Series No. 233, STI/DOC/10/233); karton. ÖS 560,–

Radioimmunologische Verfahren zur Bestimmung von Hormonen und von anderen biologisch hochwirksamen Verbindungen in sehr niedrigen Konzentrationen bewähren sich seit zwei Jahrzehnten in der klinischen Diagnostik hervorragend. Sie finden seit wenigen Jahren — mit einem nicht ganz verständlichen Zeitverzug — auch Eingang in das Gebiet der tierischen Reproduktion. Diesem Trend Rechnung tragend hat die IAEA gemeinsam mit der FAO Anstrengungen unternommen, um von einem Kreis einschlägiger Fachleute die hier vorliegende Praktikumsanleitung verfassen zu lassen. Die beim "Interregional Training Course on Radioimmunoassay and its Application in Research on Animal Reproduction" an der Cornell Universität 1982 gewonnenen Erfahrungen und die dort durchgeführten Laborversuche waren Grundlage für das Buch.

Eine ca. 50 Seiten umfassende allgemeine Einführung informiert über Radioaktivität, Zerfallsarten, Zerfallsgesetze, Wechselwirkung zwischen Strahlung und Materie sowie über Nachweismethoden und die Funktion diverser Detektoren, über Zählstatistik und nicht zuletzt über Regeln zum Strahlenschutz.

Unter „Einführung in die praktische Arbeit" — dem nächsten Kapitel — verstehen die Autoren zunächst Hinweise auf Faktoren, die mit Tracern erzielte Ergebnisse beeinflussen können: Isotopieeffekte sowie der Einfluß von Strahlung und von chemischen Verunreinigungen. Im weiteren wird in diesem Abschnitt eine Einführung in die Endokrinologie der Reproduktion von Säugetieren gegeben sowie die Funktionsweise des Radioimmunoassays erklärt.

Weitere 100 Seiten sind einigen z. T. recht ausführlichen Praktikumsaufgaben gewidmet. Ausgehend von Übungen, die u. a. die Messung der Radioaktivität, Absorption und Selbstabsorption, das LSC (liquid scintillation counting), die Löschung und Löschkorrektur zum Inhalt haben, sind Erzeugung und Charakterisierung von Antiseren sowie radioimmunologische Bestimmung einiger für die Tierproduktion wichtiger Hormone (Progesteron, Testosteron und LH) Gegenstand des Experiments.

60 Seiten Anhang enthalten u. a. Tabellen zum Strahlenschutz, die Ableitung wichtiger Formeln, eine Tabelle von Logit-Werten und schließlich ein knapp 200 Stichworte umfassenden Glossarium mit Begriffen ausschließlich Radioaktivität und Isotopenanwendung betreffend.

Die allgemeine Einführung geht etwas über das hinaus, was an Kenntnis für den Anwender eines RIA erforderlich ist. So erhält auch der, der weitergehende Tracerexperimente beabsichtigt, notwendige und nützliche Hinweise zum praktischen Handeln.

Die Versuche selbst sind so angelegt, daß der Lernende an aktuellen Beispielen mit den wichtigsten methodischen Varianten vertraut gemacht wird. Den einzelnen Abschnitten und Übungsaufgaben folgen Literaturhinweise, die die wichtigsten, auch neuere, Arbeiten für eine gezielte Vertiefung des Stoffes enthalten.

Der vorliegende Band enthält auf 250 Seiten, knapp und präzis formuliert, notwendige Grundlagen und praktische Hinweise für die RIA-Anwendung in der Tierproduktion, insbesondere der tierischen Reproduktion. Darüber hinaus findet der an einer Traceranwendung in der Biologie interessierte Leser nützliche Information und Anleitung zur praktischen Arbeit. Ein für den angesprochenen Leserkreis empfehlenswertes Buch.

G. Vormum

Advances in Dosimetry for Fast Neutrons and Heavy Charged Particles for Therapy Applications Proceedings of an Advisory Group Meeting, Vienna, 14.—18. June 1982

Vienna: IAEA 1984. 301 S., div. Abb., Tab. und Lit. (Panel proc. ser. STI/PUB/643); karton ÖS 590,00

In 18 Vorträgen werden die Probleme und Fortschritte zur Dosimetrie schneller Neutronen und schwerer geladener Teilchen für Strahlentherapiezwecke zusammengefaßt. Die Gebiete

- Dosimetrie schneller Neutronen
- Aktivitäten der Primär-Standard-Laboratorien zur Neutronendosimetrie und
- Dosimetrie schwerer geladener Teilchen in der Strahlentherapie und Strahlenbiologie

sind die Hauptthemen.

Dem Report sind Empfehlungen aus dieser Beratung vorangestellt.

Dem Anwender werden Meßverfahren zur Bestimmung der Energiedosisverteilung in homogenen und inhomogenen Medien zusammengefaßt vorgestellt. Die Erreichung einer optimalen Dosisverteilung bei Berücksichtigung der auftretenden Fehler durch Sekundärstrahlung wird untersucht. Für den Effekt ist die Charakterisierung der Strahlenqualität entscheidend, was sich insbesondere bei Mehrfeldertechnik und Bewegungsfeldern auswirkt. Berechnungsverfahren zur Dosisverteilung werden mit Meßergebnissen verglichen. Die Bemühungen zur Standard-Messung und -Verbreitung werden durch die PTW/BRD, das NBS/USA, das BIPM/Frankreich und das NPL/GB vorangetrieben und ihre Meßmethoden und -möglichkeiten vorgestellt. Verfahren zur Dosimetrie von Ionen und π-Mesonen werden vorgestellt.

Die therapeutische Anwendung der behandelten Strahlenarten sollte Einrichtungen vorbehalten sein, die über reiche wissenschaftliche und praktische Erfahrung in der

Planung und Durchführung strahlentherapeutischer Maßnahmen mit Beschleunigern und ^{60}Co-Telegeräten besitzen. Vor der Einführung dieser Therapieform mit schnellen Neutronen, Ionen und π-Mesonen sollten Kosten-Nutzen-Rechnungen angestellt werden.

Interessenten sind Strahlentherapiezentren, die sich mit der Einführung neuer Therapieverfahren befassen bzw. Hochschul- und Fortbildungseinrichtungen. *W. Wulff*

Nuclear Analytical Chemistry

D. Brune, B. Forkman, B. Persson. Lund (Schweden): Studentlitteratur 1984. 557 S., 215 Abb., 495 Lit.; geb. kr 348,—

Die Aktivierungsanalyse hat heute ein Anwendungsgebiet, das von der Bestimmung von Spurenelementen in Reinststoffen und organischen Geweben über die Kontrolle der Umweltverschmutzung und die Bestimmung von Metallen auf dem Meeresboden bis hin zur in-vivo-Analyse im menschlichen Organismus reicht. Die Autoren des vorliegenden Buches kommen aus den von ihnen selbst so bezeichneten Fachrichtungen "Nuclear Physics and Chemistry", "Nuclear Medicine" and "Radiophysics" und behandeln unter dem an sich umfassenderen Thema "Nuclear Analytical Chemistry" die Methoden der Aktivierungsanalyse, die Meß- und Aktivierungstechnik sowie eine Reihe von Anwendungen. Um den Stoff auch Wissenschaftlern anderer Fachrichtungen leichter zugänglich zu machen, werden im Teil A und B auf 324 Seiten die Grundlagen der Radioaktivität und der Strahlenmeßtechnik in verständlicher, gut gegliederter Form recht ausführlich beschrieben. Das Buch bringt dabei nicht nur das, was in anderen einschlägigen Büchern auch steht, sondern es enthält z.B. auch ein Kapitel über Aufbau und Funktionsweise von kernelektronischen Schaltungen und Mikroprozessoren. Im Teil C wird die Aktivierungsanalyse mit Neutronen, geladenen Teilchen und Photonen auf 120 Seiten dargestellt. An zahlreichen Stellen wird auf die Hauptanwendungsbereiche hingewiesen. Im Kapitel über die Aktivierungsanalyse mit schnellen Neutronen werden besonders die Sauerstoff- und Stickstoffbestimmung und im Kapitel über die Aktivierung mit geladenen Teilchen die Bestimmung von Konzentrationsprofilen, die Oberflächen- und Tiefenanalyse und die Multielementanalyse mit Hilfe von PIXE erörtert. Bei der Diskussion der an Bedeutung zunehmenden Photonenaktivierung werden auch die Methoden der Resonanzstreuung von Gammastrahlen und die photoneninduzierte Röntgenfluoreszenzanalyse besprochen.

Im letzten Teil des Buches (ca. 120 Seiten) findet man ein Kapitel über die Aufbereitung von Bestrahlungsproben und die radiochemischen Trennmethoden sowie Kapitel über Anwendung der Aktivierungstechnik in speziellen Fachgebieten: der biologischen und medizinischen Forschung, der Kriminologie, der Umweltforschung und der Geologie. Es ist vielleicht wünschenswert, daß so wichtige Anwendungsgebiete wie z.B. Halbleitertechnik, Reinststofforschung, Metallurgie u. a. auch in speziellen Abschnitten dargestellt werden.

Am Schluß des Buches findet man eine sehr nützliche Abhandlung über den Strahlenschutz und die sichere Handhabung radioaktiven Materials.

Nach jedem Kapitel folgen Literaturangaben, insgesamt 495 Zitate, von denen 11% aus den Jahren 1983—79 und 61% aus den Jahren 1978—1969 stammen.

Insgesamt ist das Werk sehr modern angelegt, mit klarer Gliederung, übersichtlichen und instruktiven Abbildungen und Tabellen. Teil A und B sind an sich schon ein ausgezeichnetes Lehrbuch über die Grundlagen der Radioaktivität und die Strahlungsmeßtechnik.

Das Buch wird einen breiten Leserkreis finden, vor allem wird es Wissenschaftlern verschiedenster Gebiete (wie analytische Chemie, Biochemie, Medizin, Landwirtschaft, Umweltschutz) Wissen und Anregungen vermitteln. *R. Otto*

Carbon Dioxide
Current Views and Developments in Energy/Climate Research
2nd Course of the International School of Climatology, Ettose, Majorana Center for Scientific Culture, Erice, Italy, July 16—26, 1982

W. Bach, A. J. Crane, A. L. Berger, A. Longhetto (Eeds.). Dordrecht — Boston — Lancaster: D. Reidel Publ. Comp. 1984. 525 S., 153 Abb., 59 Tab.; geb. Dfl. 180.—; US $ 72.00

Im Jahre 1979 wurde in Erice eine internationale Schule für Klimatologie gegründet mit dem Ziel, spezielle Seminare zu gegenwartsbezogenen Problemen auf dem Gebiet der Klimatologie abzuhalten. Der zweite Kurs dieser internationalen Schule fand im Juli 1982 statt, und das Buch berichtet über die dort gehaltenen Vorträge — insgesamt 16. Es nahmen 38 Wissenschaftler aus 9 Ländern teil.

Was vor über einhundert Jahren vielleicht als ein kreatives Ideenspiel begann, hat sich heute zu einem Problem der Beherrschung und Prognose entwickelt, worüber fast jede namhafte wissenschaftliche Zeitschrift in ihren Ausgaben zu berichten hat: das CO_2-Problem, auch als Glashausproblem hinlänglich bekannt. In Erice wurden die Problemkreise dazu diskutiert: der Kohlenstoff-Kreislauf, die klimatischen Effekte und Tragweite und Strategien. Nach wie vor stehen die Fragen zum präindustriellen CO_2-Pegel der Atmosphäre, die Modellierung des CO_2/Meerwasser-Austausches und die Rolle der Biosphäre im Vordergrund. Erst daraus lassen sich dann mehr oder weniger zuverlässige Prognosen über den weiteren Anstieg des CO_2-Pegels und seine Folgen ableiten. Kritisch wurden von *C. Lorius* u. *D. Raynaud* (Grenoble) die Baumring- und Eiskern-Daten des CO_2 aus der Vergangenheit diskutiert, wie beispielsweise Diskrepanzen in den $\delta^{13}C$-Werten von Baumringen der Nord- und Südhemisphäre, wobei die Möglichkeit einer Fraktionierung zwischen Blattwerk und Holz angedeutet wurde. Der schwierig in die Modelle einzubeziehenden Biosphäre wird durch laufende Verfeinerung der *pools* zu begegnen versucht; einbezogene Aufwärts- und Abwärtsströmungen in den Ozeanen sind weitere Verbesserungen. Wenn es gelänge, die vom CO_2 bekannten jahreszeitlichen Schwankungen auch beim O_2 nachzuweisen, wäre eine wesentlich bessere Charakterisierung der Biosphäre möglich; denn die Photosynthese verbraucht einerseits CO_2, bildet aber andererseits im stöchiometrischen Verhältnis gleichzeitig O_2. Allerdings müßten die Schwankungen im Bereich einer relativen Änderung von 10^{-7} pro Jahr nachgewiesen werden — kein leichtes Unterfangen. Was die prognostischen Einschätzungen des CO_2-Pegels betrifft, so betrachtet man mit erhobener Augenbraue am längsten eine graphische Darstellung von *W. Bach* (Münster), wonach sich bei konstantem Fossilkohlenstoffverbrauch bis 1985 und einer nachfolgenden Abnahme von jährlich 0,175 TW (1975: 8 TW) eine drastische Verringerung des CO_2-Gehaltes ab ungefähr 2010 einstellt; zumindest steigt der CO_2-Gehalt in den folgenden Jahrzehnten nie über die 400 ppm-Marke. Wird dies im globalen Maßstab gelingen? Wohl kaum, obwohl die Kernenergie und ein sparsamer und vernünftiger Umgang mit unserem Fossilkohlenstoff im Bereich des Möglichen liegen.

Daß Proceedings eine hohe Aktualität an verarbeiteter Literatur haben, ist hinlänglich bekannt, so auch im vorliegenden Fall: insgesamt 710 Literaturhinweise, allein 80% aus den Jahren 1976—1983 mit einem Maximum von 129 Arbeiten aus dem Jahre 1981 (≙ 18,2%). Die einzige Schwäche des Buches: Manche Abbildungen sind durch das angewendete Verkleinerungsverfahren bei der Herstellung schwierig zu lesen. Der vorliegende Band dürfte in den nächsten Jahren selbst häufig in Literaturzitaten künftiger Arbeiten zu finden sein, eine weite Verbreitung ist zu wünschen und wohl auch zu erwarten. *H. Hübner*

The Fokker-Planck Equation
Methods of Solution and Applications

H. Risken. Berlin — Heidelberg — New York — Tokyo: Springer 1984. 495 S., 95 Abb., 3 Tab., 324 Lit. (Springer Series in Synergetics, Vol. 18); geb. DM 125,—

Die *Fokker-Planck*-Gleichung, ursprünglich von ihren Namensträgern am Problem der *Brown*schen Teilchenbewegung entwickelt, bedeutet heute eine Gleichung, die in so unterschiedlichen Wissenszweigen wie etwa der Flüssigkeiten- und Festkörperphysik, Plasmaphysik, Laserphysik, Elektrotechnik und chemischen Physik anzutreffen ist. Nicht zuletzt diese interdisziplinären Möglichkeiten sprachen für die Aufnahme der vorliegenden Monographie von *H. Risken* in eine Schriftenreihe über Synergetik, also dem Forschungsgebiet, das sich mit den strukturbildenden (kooperativen) Phänomenen von Subsystemen eines größeren Hauptsystems befaßt.

Kap. 1 führt — gewissermaßen in einem ersten Anlauf — in die verschiedenen Darstellungsformen der *Fokker-Planck*-Gleichung ein und vermittelt ihren Zweck: Berechnung einer Verteilungsfunktion, um über geeignete Mittelwerte (Momente) makroskopische Eigenschaften beschreiben zu können. Kap. 2 stellt dann das notwendige Rüstzeug für die wahrscheinlichkeitstheoretischen Untersuchungen der folgenden Kapitel zusammen; ebenso als Vorarbeit für weitere Studien und Vergleiche kann das Kap. 3 über *Langevin*-Gleichungen angesehen werden. Kap. 4 (*Fokker-Planck*-Gleichung) schließt mit der Vorstellung der *Kramers-Moyal*-Entwicklung die Fragen zur mathematisch fundierten Herleitung der *Fokker-Planck*-Gleichung und damit gewissermaßen den ersten Teil des Buches (etwa 100 S.) ab.

Der zweite Teil der Monographie (Kap. 5 bis 10) ist ausschließlich den Lösungsmethoden der *Fokker-Planck*-Gleichung gewidmet (180 Seiten). In systematisch aufbauender Weise werden behandelt: Methoden zur Lösung der *Fokker-Planck*-Gleichung mit einer, anschließend mit mehreren Variablen, das Problem der Reduktion der Zahl der Variablen (dazwischen eingestreut ein mit Anwendungsbeispielen versehenes Kapitel über Korrelationsfunktionen und linear-response-Theorie), Lösungen über tridiagonale Rekursionsformeln sowie Anwendungen auf gewöhnliche und partielle Differentialgleichungen. Methoden zur Behandlung der *Kramers*-Gleichung (spezielle Form der *Fokker-Planck*-Gleichung zur Beschreibung der *Brown*schen Bewegung in einem Potential) runden die Betrachtungen zur Lösungstheorie der *Fokker-Planck*-Gleichung ab.

Der dritte Teil (Kap. 11 und 12) enthält auf etwa 140 Seiten Anwendungen der *Fokker-Planck*-Gleichung auf die *Brown*sche Bewegung in periodischen Potentialen (100 Seiten) bzw. die Photonenstatistik von Lasern (40 Seiten). Obwohl einerseits verständlicherweise viele Argumente für eine gründliche Behandlung von nur wenigen ausgewählten Beispielen sprechen, hätten andererseits ein oder zwei weitere deutlich herausgehobene Anwendungsbeispiele (z.B. *Fokker-Planck*-Form der *Boltzmann*-Gleichung in der Plasmaphysik) die vielseitigen theoretischen Einsatzmöglichkeiten der Gleichung auch in diesem Applikationsteil des Buches gut demonstrieren können.

Die sehr klar und übersichtlich geschriebene moderne Monographie kann neben den Spezialisten auch allen mathematisch vorgebildeten Lesern (Physikern, Chemikern, Ingenieuren, fortgeschrittenen Studenten) empfohlen werden, die auf Forschungs- und Entwicklungsgebieten arbeiten, wo sich der Einsatz dieser interessanten Gleichung lohnt. *W. Stiller*

Kinetic Theories and the Boltzmann Equation

Hrsg.: C. Cercignani. Berlin — Heidelberg — New York, — Tokyo: Springer 1984. 248 S., 6 Abb., 1 Tab., 174 Lit. (Lecture Notes in Mathematics 1048); kart. DM 33,50; US $ 12.50

Das Buch enthält den Lehrstoff von drei von vier Vortragszyklen sowie einigen Seminaren, die während einer Sommerschule vom 10. bis 18. Juni 1981 in Montecatini (Italien) zum obigen Thema von namhaften Autoren dargeboten wurden. Vom Herausgeber wird im Vorwort bedauert, daß der Inhalt des vierten Vortragszyklus (*H. Grad*: Singuläre Grenzen der *Boltzmann*-Gleichung) nicht zum Abdruck kommen konnte.

Von *J. Hejtmanek* (Univ. Wien) wird unter dem Titel „Zeitabhängige lineare Transporttheorie" (S. 1—59) das Verhalten von Testteilchen T mit unbekannter Verteilungsfunktion f_T in einem Wärmebad von Teilchen A mit fixierter Verteilungsfunktion f_A behandelt. Einbezogen werden elastische Streuungen der T- and den A-Teilchen sowie T-A-Stöße mit Absorption oder Produktion von T-Teilchen. Für zwei Modellsituationen (Neutronentransport in der Reaktorphysik, Photonentransport in der Computer-Tomographie) erfolgt dann eine detailliertere theoretische Behandlung.

H. Neunzert (Univ. Kaiserslautern) gibt eine „Einführung in die nichtlineare *Boltzmann-Vlasov*-Gleichung" (S. 60 bis 110), d. h. in das kinetische Problem der Wechselwirkung von N identischen Teilchen, wobei diese Wechselwirkung hinreichend „schwach" sein muß (z.B. Gravitation, *Coulomb*-Wechselwirkung, genauer: Modifikation dieser Kräfte). Wichtige Aspekte der *Neunzert*schen Vorlesungen sind: modifizierte *Vlasov*-Gleichung, *Vlasov*-Gleichung und vollständiger Satz der *Maxwell*-Gleichungen (um den plasmaphysikalisch wichtigen Fall eines selbstkonsistenten Magnetfeldes, erzeugt durch die sich bewegenden N Ladungen, erfassen zu können) sowie numerische Lösungsmethoden der *Vlasov*-Gleichung.

Der dritte Vorlesungszyklus, gehalten von *P. F. Zweifel* (Virginia Polytechnic Inst.) behandelt die „*Boltzmann*-Gleichung und ihre Eigenschaften" (S. 111–175). Im einzelnen enthält auch dieser durch Übersichtlichkeit und mathematische Strenge sehr bestechende Beitrag eine heuristische „Ableitung" dieses Prototyps einer kinetischen Gleichung, das H-Theorem und die Stoßinvarianten, Methoden zur Lösung der räumlich homogenen *Boltzmann*-Gleichung, eine Aufzeichnung der Schwierigkeiten bei der räumlich inhomogenen *Boltzmann*-Gleichung (beschrieben mit dem Formalismus der Quantenfeldtheorie), die Behandlung einer mathematisch vereinfachten *Boltzmann*-Gleichung ("mollified" *Boltzmann* equation) sowie die Lösungstheorie für Systeme in Gleichgewichtsnähe.

Die sich anschließenden fünf Seminartexte (S. 176–243) behandeln erste Ergebnisse über die Nichtexistenz von Lösungen für eine linearisierte *Boltzmann*-Gleichung mit drei behandelten Freiheitsgraden (*M. D. Arthur*, Univ. Ulm), die räumlich homogene *Boltzmann*-Gleichung für Molekülkräfte von unendlicher Reichweite (*T. Elmroth*, Chalmers TH Göteborg), neuere Resultate des *Cauchy*schen Problems der *Boltzmann*-Gleichung (*H. Spohn*, Univ. München) sowie spezielle nichtlineare Probleme der kinetischen Gastheorie (*T. Ytrehus*, Univ. Illinois, Urbana).

Wie auch bei den übrigen Büchern dieser den Fachmann ansprechenden Serie handelt es sich bei Bd. 1048 um Beiträge, die auf höchstem internationalem Niveau sowohl die Grundlagen als auch ausgewählte Anwendungen zur Thematik bringen. Zum Leserkreis dürften deshalb vor allem Mathematiker und theoretische Physiker zu zählen sein, ferner mathematisch versierte Ingenieure, die auf dem Gebiet der Gasdynamik arbeiten. *W. Stiller*

Matrizen und ihre Anwendungen
Teil 1: Grundlagen

R. Zurmühl, S. Falk. Berlin — Heidelberg — New York — Tokyo: Springer 1984, 342 S., 53 Abb.; geb. DM 84,—; US $ 33.00

Das vorliegende Buch ist bereits die fünfte Auflage dieses weitverbreiteten Standardwerkes über eine der wichtigsten Disziplinen der angewandten Mathematik. Bis zur vierten Auflage erschien es unter dem Titel „Matrizen und ihre technischen Anwendungen". Seither sind etwa 20 Jahre vergangen, und es war damit eine starke Überarbeitung und Erweiterung zu erwarten. Das auffälligste Merkmal der fünften Auflage ist die Trennung in zwei Bände (Teil 1: Grundlagen, Teil 2: Numerische Methoden und technische Anwendungen).

Der hier vorliegende Teil 1 umfaßt 6 Kapitel (Kap. 1 — Matrizenkalkül, Kap. 2 — lineare Gleichungen, Kap. 3 — quadratische Formen nebst Anwendungen, Kap. 4 — Eigenwertaufgaben, Kap. 5 — Struktur der Matrix, Kap. 6 — Blockmatrizen), in denen die Matrizenrechnung systematisch aufgebaut wird. Die theoretischen Grundlagen und Sätze sind ausführlich dargelegt und durch zahlreiche Beispiele veranschaulicht. Das Buch beinhaltet zum größten Teil den klassischen Stoff der Matrizenrechnung. Neu gegenüber den bisherigen Auflagen sind Abschnitte über Parametermatrizen (Kap. 5), Blockmatrizen (Kap. 6) und die Expansion von Polynommatrizen (Kap. 6). Bei der Expansion von Polynommatrizen sind zwei Verfahren von herausragendem theoretischem und praktischem Interesse, wobei eines von beiden (die Diagonalexpansion) von einem der Verfasser stammt. Der Zugang zur weiterführenden Literatur wird durch zahlreiche als Fußnoten in den Text eingestreute Literaturhinweise hergestellt.

Wegen seiner praxisnahen und klaren Sprache bei gleichzeitig präziser Darstellung wird dieses Buch seine Bedeutung als Standardwerk für Ingenieure der verschiedensten Bereiche sicher weiter ausbauen. Auch für Naturwissenschaftler, die sich mit der Anwendung der Mathematik in ihren Spezialgebieten beschäftigen, stellt dieses Buch ein wertvolles Hilfsmittel dar. *G. Kummer*

Einführung in die Programmierung mit PASCAL

H. E. Erbs, O. Stolz. 2. erw. Aufl. Stuttgart: B. G. Teubner 1984. 240 S., zahlr. Abb. u. Tab. (MikroComputer-Praxis); kart. DM 24,80

Sicher nicht zuletzt wegen der — leider unüblichen — humorigen aber doch anspruchsvollen Darstellung erschien innerhalb von zwei Jahren bereits die zweite Auflage dieses Buches. Sie stellt eine etwas erweiterte Fassung der ersten Auflage (Rezens. siehe Isotopenpraxis H. 11/1983) dieses mit glücklicher Hand geschriebenen „Lehrbuches" des Programmierens (speziell mit PASCAL) dar. Die Erweiterung berücksichtigt hauptsächlich eine in der Zwischenzeit erschienene internationale Normierung der Sprache PASCAL. Dadurch können zwar jetzt beliebig (endlich) dimensionierte Reihen in Prozeduren formuliert werden, eine dynamische Dimensionierung ist aber auch hier von der Anlage der Sprache her nicht möglich.

Auch dieser Auflage wünscht man eine weite beispielgebende Verbreitung. *R. Straubel*

ISOTOPENPRAXIS

Band 21 — 1985 — Heft 8

Contents / Inhalt

►Language of Publication — ►Publikationssprache

Isotopenpraxis is indexed in Current Contents/PC & ES, Science Citation Index, and INIS Atomindex

FSC
www.fsc.org
MIX
Papier | Fördert gute Waldnutzung
FSC® C083411

Kolibri 360 GmbH
Ferdinand-Jühlke-Straße 7
99095 Erfurt, Deutschland
produktsicherheit@kolibri360.de